THE SMARTEST
ANIMALS
ON THE PLANET

THE SMARTEST
ANIMALS
ON THE PLANET

EXTRAORDINARY TALES OF THE NATURAL
WORLD'S CLEVEREST CREATURES

SALLY BOYSEN

**WITH A CONTRIBUTION FROM
DEBORAH CUSTANCE**

Published in 2009 by
A&C Black Publishers
36 Soho Square
London W1D 3QY
www.acblack.com

ISBN 978-1-4081-1343-1

Conceived, designed and produced by
Quarto Publishing plc
The Old Brewery
6 Blundell Street
London
N7 9BH

Senior editor: Lindsay Kaubi
Text editors: Claire Waite Brown and Richard Rosenfeld
Art director: Caroline Guest
Managing art editor: Anna Plucinska
Designer: John Grain
Design assistant: Saffron Stocker
Illustrators: William Donohoe and Malcolm Swanston
Picture researcher: Sarah Bell

Creative director: Moira Clinch
Publisher: Paul Carslake

Colour separation by SC(Sang Choy) International
Pte Ltd, Singapore
Printed in Singapore by Star Standard Industries Pte Ltd

9 8 7 6 5 4 3 2 1

CONTENTS

FOREWORD

My interest in animal intelligence grew from my love of animals when I was in elementary school. A voracious reader, I devoured every book about animals at our local library, and even at a very young age, my goal was to become a veterinarian for zoo animals. My particular love was for the great apes – especially orang-utans and chimpanzees. However, my first college course in ethology, or animal behaviour, rather than animal medicine, changed everything. I soon realized that what animals were doing was much more exciting to learn about. The particular research direction that I chose, the study of chimpanzee cognitive abilities, has included experiments designed to explore chimpanzee numerical skills, their understanding of scale models and the ability to demonstrate an understanding of causality when using tools.

Even after 35 years of working with chimpanzees, their minds and behaviour still hold the same fascination for me that they did when I was 7. My hope for this book is that it will arouse similar excitement in the reader and pique their curiosity about many animal species, each with their own unique capacities that reflect a measure of intelligence. For some animals, this will be easy to see simply by using an understanding of your own behaviour and skills for comparison. More challenging may be the abilities of more distant and less well-known species that require more thoughtful consideration of how their natural habitat may have contributed to specialized behaviours that are "smart" for their particular environment. Consider what you read to be merely a drop of water in a sea of newly discovered knowledge and remarkable findings about all the other smart animals on the planet!

Dr Sally Boysen

ABOUT THIS BOOK

This book is organized into seven chapters: Using tools, Communication, Imitation and Social Learning, Mirror Self-recognition, Numerical Abilities, Animal Language Studies and Cooperation and Altruism. Each chapter looks at the various species of animals that have shown particular skills and talents in these specific areas and describes the processes researchers use to observe and detect these signs of intelligence.

INTRODUCTION

In the introduction, learn how scientists define and search for intelligence in animals.

FASCINATING FACTS

Interesting facts about the animals in the book provide background information on how they live.

MAPS

A map on each article shows where in the world each animal lives in the wild.

STEP-BY-STEP DIAGRAMS

Step-by-step diagrams demonstrate in detail the behaviours of smart animals and the experiments used to test them.

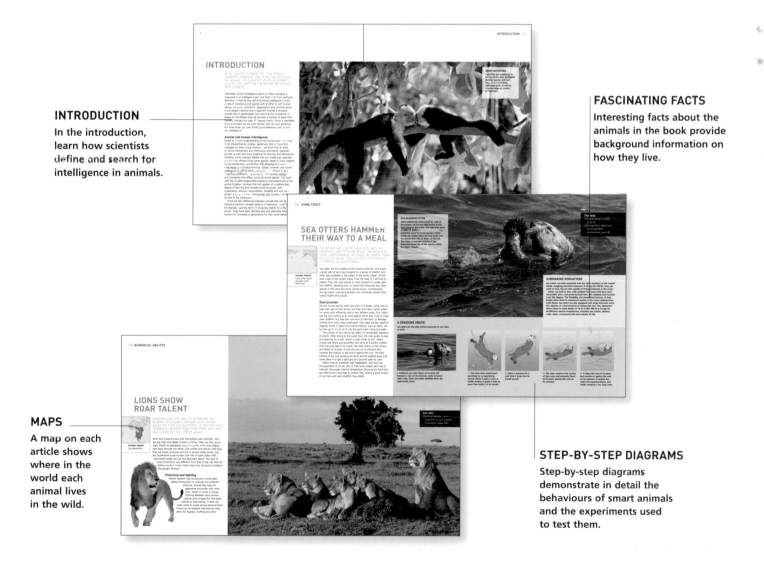

INTRODUCTION

WHAT EXACTLY IS MEANT BY "THE WORLD'S SMARTEST ANIMALS" AND HOW CAN WE JUDGE AN ANIMAL'S INTELLIGENCE WHEN ACADEMICS CAN'T EVEN AGREE ON THE NATURE OF HUMAN INTELLIGENCE?

Ultimately, human intelligence seems to reflect whatever is measured in an intelligence test, and that's a far from satisfying definition. It may be that defining animal intelligence is more a case of comparing one species with another or with human beings, using our experience, observations and common sense. It just appears obvious that a capuchin monkey is probably smarter than a grasshopper, but note that the comparison is based on the abilities that we perceive a monkey to share with humans. Perhaps this type of "species-centric" focus is inevitable, since as humans we are most familiar with our own behaviour. But what drives our own thinking and behaviour and, in turn, our intelligence?

Animal and human intelligence

Based on current understanding of the human brain – a 1.4 kg (3 lb) extraordinarily complex, gelatinous blob of tissue that manages our every living moment – we know that an array of neural mechanisms and information-processing capacities provide us with enormous potential for learning and behavioural flexibility. Some scientists believe that our intellectual capacities are distinctly different from other species, based in many respects on the tremendous contribution that language and culture have made on individual learning. Others, however, see human intelligence as part of what Charles Darwin referred to as a "cognitive continuum", a distribution of cognitive abilities and complexity that differs across all animal species. This starts with the simplest single-celled organism, and extends across the animal kingdom. Animals that are capable of a sophisticated degree of learning and complex social structures, with cooperation, altruism, reconciliation, empathy and tool use – as seen among dolphins, chimpanzees and humans – lie at the far end of the continuum.

There are also differences between animals that can be readily trained to perform complex patterns of behaviour – with dogs, for example, opening doors or retrieving objects for a disabled person. Dogs have been domesticated and selectively bred by humans for hundreds of generations for their social behaviour

KEEN CAPUCHINS
Capuchins are considered to be one of the most intelligent monkey species and have been observed making and using tools, as well as showing signs of a sense of "fairness".

and learning capacities. But this exceptional ability to learn, and an accompanying high degree of sociability with humans, can make them appear quite intelligent when, in fact, it is the combination of their willingness to train and receive rewards, coupled with their genetically shaped willingness to please a trainer, that results in such accomplishments.

Great apes

The great apes are closest to humans in terms of problem-solving abilities and cognitive flexibility, with the chimpanzee being the closest living primate species to our own. It shares over 98 percent of its genetic material, or DNA, with humans. Because we shared a common ancestor a mere five or six million years ago – quite a short period in the evolutionary and geologic timetable – it should be no surprise that we share a tremendous overlap in features. These include anatomical, physiological, morphological, neurological and behavioural similarities, with the capacity to learn, and some types of problem-solving that are specific to the ape lineage. For example, the propensity for tool manufacturing and tool use can be observed in both humans and chimpanzees, but while our own culture has become incredibly sophisticated at both, the use and construction of different toolkits among various chimp communities across Central and Western equatorial Africa is still at a very basic level.

Comparative cognition

We have selected topics that cut a wide swathe across the developing field of animal or comparative cognition. This covers a wide diversity of species, methods, topics and scientific questions that are providing glimpses into the possible minds and intelligence of other animals. The present chapters can only touch upon a portion of the many extraordinary approaches and questions that are being addressed, and only some of the significant pioneering studies that have propelled the subject forward as an important new discipline.

Given our own history of tool-making and tool use, it should not be surprising that comparisons with other animals' use of objects, particularly for acquiring food, is of great interest to researchers in many disciplines. Similarly, because of the enormous scientific interest in the evolution of human language, studies of animal communication and vocalizations that may show evidence of referential signal use is of enormous interest to anthropologists, linguists, psychologists, biologists and philosophers.

In the same vein, the remarkable studies of sign language and other artificial language systems taught to a number of chimpanzees, one orang-utan and one gorilla over the past 40 years, have revealed much about the similarities for representational symbol use and comprehension that exist between those species and our own. In turn, the experimental research on mirror self-recognition in apes, dolphins, elephants and now even magpies continues to challenge our understanding of other minds, and the parameters of what it means to be a chimpanzee or a dolphin, for example, in terms of a self-concept. These and other key questions that have been, and will be, explored in other animals challenge our understanding of what it means to be an aware, conscious, sentient being. We also need to know to what extent other species share these characteristics. Perhaps only then can we become proper stewards for all the species along the Darwinian continuum.

NOT SO BIRDBRAINED
Parrots are known for mimicking human speech but studies on African grey parrots have shown that they can actually use and understand human language.

SMART SWIMMERS
Dolphins have a reputation for being intelligent, and display many "smart" behaviours such as recognizing themselves in mirrors.

CHAPTER ONE
USING TOOLS

For a long time it was assumed that humans were the only species to make and use tools. As the following pages show, nothing could be further from the truth. More and more evidence is being documented of the ingenuity with which wild and captive animals fashion and use materials in their habitats as tools to enhance their strength, reach, stability and other physical abilities in the struggle to survive. Marine mammals, birds, monkeys, elephants and, of course, great apes all adapt and use tools to forage, hunt and protect themselves. Some of the behaviours have clearly been learned, some appear to be instinctive, but all are amazing.

WOODPECKER FINCHES PROBE WITH STICKS

WOODPECKER FINCHES ON SANTA CRUZ IN THE GALÁPAGOS ISLANDS USE STICKS AND CACTUS SPINES AS PROBING TOOLS IN THE MORE ARID CLIMATES ON THE ISLAND, COMPARED TO FINCHES LIVING IN EVERGREEN FORESTS THAT HAVE ABUNDANT FOOD.

NATURAL HABITAT
Galápagos Islands,
Southwest Pacific

Although there is presently a great deal of information about tool use across a variety of animal species, few studies have provided substantial evidence for the ecological importance of tool use. Observational studies of one species of bird that lives on Santa Cruz in the Galápagos Islands, the woodpecker finch, may be the exception. Other birds have been observed to use tools on occasion, including Egyptian vultures, green-backed herons, satin bower birds and New Caledonian crows; however, it was thought that tool use by the woodpecker finch, one of the 13 remaining species of the original 15 studied by Charles Darwin during the voyage of the HMS *Beagle*, evolved in response to areas on Santa Cruz where there were dry and unpredictable habitats.

A tale of two climates

Woodpecker finches have been seen using twigs or cactus spines to pry insects out of crevices and tree cavities. Use of these tools allows them to reach particularly large, inaccessible prey that is hidden from view. Scientists studying other animals have discovered that tools allow their subjects to access and extract food from non-visible locations, enhancing their diet. To explore the potential contribution that tool use might have for the woodpecker finches, a research team compared different populations of finches that lived in two different climate zones, and also compared each of the two climates during

the wet season and the dry season. They theorized that the two groups of finches would display different amounts of tool use as a function of the two seasons and the two climates.

The first climate was the Arid Zone, located near the coast and consisting of semi-desert, open-canopy forest, with two types of cactus and deciduous trees and bushes. The second climate was the Scalesia Zone, an evergreen forest at a higher elevation that was a lush cloud forest of evergreen trees and mosses, and was usually wet all year round.

In the Arid Zone, food was more limited, so two predictions emerged: 1) finches should display more tool use in the Arid Zone; and 2) there should be more tool use observed during the dry season, when resources would be even more limited. Because the birds were quite used to human tourists being on the island, observations by the research team could be made at quite close range, less than 10 m (30 ft), therefore allowing for prey items to be identified. Judgements of prey size were made, relative to the average known beak size of woodpecker finches.

Tool use was defined as holding a twig or cactus spine in the beak and inserting it into a crevice or opening in the tree bark. In addition, several behaviours were identified and considered to be probe actions by the birds. These included: 1) inserting the beak into moss or curled leaves; 2) removing bark using the

PROBING FOR FOOD
Woodpecker finches use twigs or cactus spines, available all around the arid climate areas on Santa Cruz, that allow them to probe into crevices or among leaves, lichens, moss and in the bark of dead or live trees to find insects and other prey.

See also
Use of spears by wild chimpanzees, *page 38*

beak, labelled as "chip off"; 3) "glean", defined as taking prey directly from the surface of the substrate; 4) "peck", a self-explanatory behaviour meaning to peck at, using the beak; and 5) "bite", when a bird bit into the surface of leaves. These measurements were taken when the birds foraged on moss, leaves, the bark of dead wood and lichens.

Choose your tool wisely

The finches turned out to use different foraging substrates between the two climate zones, as well as between the dry and wet seasons in each zone. For the most part, they foraged in moss in the Scalesia Zone during the wet season, but used bark and leaves in the dry season. When finches foraged in the Arid Zone, they chose bark first – even though bark there was much more difficult to remove than in the Scalesia Zone – and tree holes second. However, during the wet season in the Arid Zone, the birds foraged through bark, but also lichens, leaves and fruit, and searched in tree cavities. An evaluation of the amount of food extracted through tool use revealed that fully one-half of prey was obtained through extractive foraging by finches living in the Arid Zone. These results provided strong evidence for the ecological importance of tool use by woodpecker finches living in climates where food is scarce, and the significant contribution to their diet that is made by extractive foraging.

NEW CALEDONIAN CROWS HOOK A TREAT

NATURAL HABITAT
New Caledonia,
Southwest Pacific

IT HAS BEEN ASSUMED PREVIOUSLY THAT ONLY NON-HUMAN PRIMATES USE MULTIPLE TYPES OF TOOLS HABITUALLY. HOWEVER, THE CROWS OF NEW CALEDONIA MUST NOW BE INCLUDED IN THE LIST OF ANIMALS THAT USE A RANGE OF TOOL TYPES, SINCE THEY CARVE OUT STICKS, TWIGS AND PORTIONS OF LEAVES TO HELP THEM OBTAIN FOOD.

Members of the corvid family of birds include wild crows, magpies, jackdaws, rooks and ravens, all of whom have long been known for demonstrating creative behaviours indicative of their high intelligence. Field scientists have recently made systematic observations of crows that are endemic to New Caledonia, an archipelago located some 1,500 km (930 miles) from Australia, where tool use by crows is part of the native folklore. Scientists have discovered a remarkable array of tools used by these birds, which include twigs, portions of leaves from the pandanus plant, moulted crow feathers, leaf stems and even pieces of salvaged cardboard.

Design and operation

The crows use tools from three categories: straight sticks or leaf stems; hooked twigs or vines; and torn pieces from the leaves of the pandanus tree. Because the shape of the leaf tools from the pandanus tree appears to vary across New Caledonia, there may be local variations that are represented in certain communities of crows. This suggests that the techniques may be transmitted among crows in the group through imitation and observational learning, and may represent a primitive type of protoculture.

The stick, hook and leaf tools that are manufactured by the crows are used to extract grubs from inside logs and trees, and recent observations have provided detailed descriptions of how they manufacture the hooked-twig and stepped-cut tools. Hooked-twig tools are broken from a larger twig and taken to a tree so the crow can work on the small hook at the end with its beak. The bark is then stripped off, followed by removal of the leaves. Stepped-cut tools are cut along the edges of pandanus tree leaves to create a tool with a tapered, pointed shape, with barbs along the uncut edge, facing upwards from the narrower end.

HOW OTHER BIRDS USE TOOLS

Only a few other bird species have displayed a similar type of tool use: the woodpecker finch uses stick tools (see page 14); hyacinth macaws use wood slivers or leaves for wedging nuts so they can crack them more efficiently; and Egyptian eagles drop stones on ostrich eggs.

See also
The wild chimp's toolkit,
page 34

INNATE BEHAVIOUR
Recent studies of captive-reared crows
show they begin to use tools spontaneously
at about two and a half months of age,
without observing adult crows using tools
or seeing demonstrations by humans.

MADE TO ORDER

New Caledonian crows modify a variety of
materials to make effective tools that they
use as probes while searching for grubs in
dead logs or live trees. Tools are often reused
at new sites.

   Wild crows are unique among bird species
in the variety and systematic ways that they
modify twigs and leaves to create more
functional tools.

One of the most remarkable and resourceful
of creatures, the New Caledonian crow is
celebrated on a local postage stamp.

A pandanus tree
leaf is cut to create
a tapered tool with
barbs along the
uncut edge.

Twigs are broken off and
then stripped of bark.

The barbs from long feathers
are removed, and the quill
and shaft used for probing.

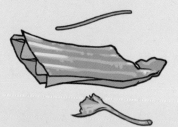

Leaf stems and
cardboard tools.

Crows use tools in both dead and live trees, and will search in numerous sites in the same tree. Their tools can be operated in a variety of ways: rapid movements are used when the prey is buried under leaves, whereas slow movements are preferred when trying to hook a food source in a small hole, or one that is not visible prior to capture.

The crows often take their tools with them to a new site, and tuck them under their feet during feeding so the tool can be used again. What is truly exciting is the discovery that New Caledonian crows have exceeded the skills demonstrated by all other non-human species, because they seem to recognize the functional requirements of a tool for a specific search effort, and are also the first species to use hooks as tools. In addition, tool types are similar enough across different groups to suggest a kind of standardization of tool types, with the crows manufacturing specific tools for different functions.

Studies in captivity

Observing crows in the wild is extremely difficult, so in order to address a number of questions about how crows develop the use of tools, studies of captive crows have recently been conducted (see page 20). To investigate the crows' abilities more closely, captive studies have documented the development of tool use by naive, hand-reared crows. In one study, two juvenile crows were exposed daily to a human tutor who demonstrated the use of a stick tool to them by extracting food from baited holes and crevices. The tutor only demonstrated how to use the tool, but never how to make a tool. Two other juvenile crows experienced the same amount of human interaction in the same type of housing environment, but never witnessed any tool use by a human tutor. However, when all the crows were 79 days old, all four birds were using stick tools to retrieve baited food from crevices that had been built in their aviary homes. Their use of tools had developed spontaneously. There was no difference in the onset of tool use by any of the birds, suggesting that some features related to tool use by crows may have a genetic basis.

From their observations, scientists determined that before the crows were proficient at using twig tools they engaged in repetitive behaviours with twigs, in a kind of "pretend probing", holding the twig in their beaks and moving their heads back and forth against their perches. The whole series of movements was highly similar to how adult crows were observed using stick tools in the wild. This type of practice is necessary for the crows to perfect their tool-using skills, much like someone would practise a tennis serve or golf swing. It appears that certain behavioural predispositions may be inherited for tool use in crows, and birds perfect their skills through social learning and experience.

FAMILY LIFE

New Caledonian crows live in family groups and, as omnivores, eat many different types of food, including numerous species of insects, invertebrates, eggs, nestlings, snails that are dropped from heights onto rocks and several types of nuts and seeds.

CAPTIVE CROWS SHOW TALENT FOR TOOL USE

NATURAL HABITAT
New Caledonia,
Southwest Pacific

CREATIVE EXPERIMENTS WITH CAPTIVE NEW CALEDONIAN CROWS HAVE EXPANDED OUR UNDERSTANDING OF THEIR ABILITY TO USE TOOLS, AND DEMONSTRATED THEIR ABILITY TO USE TOOLS TO SOLVE SPECIFIC PROBLEMS.

Following the reports of innovative tool use by New Caledonian crows in their natural habitat, scientists were eager to explore these abilities further under experimentally controlled conditions. Using captive-bred and captive-raised crows, a series of unique experiments were devised that have revealed a greater depth and flexibility in the New Caledonian crow's tool-using capacity.

Previous observations in the wild have provided documentation that these birds are capable of using different types of tools to exploit different food sources. However, studies completed with captive birds have allowed for more rigorous testing using unique tasks and materials that wild crows would never encounter. The skill levels revealed in these experiments indicate that New Caledonian crows have capacities that are unparallelled when compared with other bird species. Only the great apes – especially chimpanzees – and humans have shown similar flexibility and ingenuity in the use of tools.

A research group at Oxford University was keen to find out if the crows understood the relationships inherent in tasks that require tool use for solution. Do New Caledonian crows really understand what is needed to retrieve a food reward in a particular experiment? Is it possible to create a test situation that will reveal something more about how the birds are

CAPACITY FOR LEARNING
The studies carried out on New Caledonian crows in captivity proves that tool-making and successful tool use developed spontaneously.

See also
New Caledonian crows hook
   a treat, *page 16*
The wild chimp's toolkit,
   page 34

reasoning (if they are, indeed, reasoning at all) in order to fetch the tasty food reward that the experimenters have placed out of reach? The group devised a series of exciting studies that led them to reach a number of conclusions.

Inherent behaviour

A natural starting point for trying to understand tool use in New Caledonian crows was to study the development of the skills over time in very young birds. It was important that the young crows did not have the opportunity to see any other crows using tools, in order to observe how the birds respond to available tool materials for the very first time on their own. A further test of how the crows learn to use tools was also included in the experimental design. Half of the young birds never saw any type of tool use by other crows or their human care-givers, and the other half of the subjects had daily demonstrations from their human foster parent, showing them the correct way to use twigs as probes. Despite these differences in early experiences, all of the juvenile crows showed the ability to use twigs as probes by the time they were about three months old. These results support the idea that tool use in New Caledonian crows develops spontaneously. In addition, the birds all showed earlier components of tool use, such as holding a twig in their beaks and moving their heads back and forth, clearly behaviours that are necessary for successful tool use. These observations, together with the emergence of tool-using abilities by all the juveniles, coupled with the frequent tool use seen in wild New Caledonian crows, suggests that the birds are born with behavioural predispositions for highly similar movements that support using tools.

Solving a problem with tools

Scientists were also interested to find out whether or not the crows could use tools effectively for specific problem-solving tasks. By changing the type of task and materials available, the researchers could learn whether the crows had flexibility with their tool-using capacities. To test these ideas, captive-born adult New Caledonian crows were presented individually with tasks that required the use of a specific tool type in order to obtain a food reward that was out of reach. For example, the birds readily used sticks or other probe tools to extract food from crevices or holes prepared by the experimenters, replicating the types of natural use of twigs seen in wild crows. When more challenging problems were presented, the birds were again successful and a crow named Betty was even observed bending wire to fashion a hook with which she could retrieve a tiny pail of food from a clear plastic tube. While this type of tool-making has only been reproduced under laboratory conditions, it does suggest that New Caledonian crows are able to use flexible behaviours that may result from reasoning about a specific problem and the necessary requirements for its solution.

THE STICK AND TUBE PUZZLE

The stick and tube puzzle was originally designed to test for tool use in primates. Only the great apes and capuchin monkeys have consistently solved this puzzle. No-one expected that a bird would be able to solve it, but Betty has managed it.

1 | A piece of food is placed halfway down the length of a transparent horizontal tube. In the foreground is a probing stick.

BETTY'S INVENTION

Betty, an adult female crow, retrieves a pail of food from the bottom of a clear tube. She uses a piece of wire that she fashioned into a crude hook by inserting it in a hole in the wall of her room and pushing it in one direction. She was not trained to do this by her human care-givers, but invented the technique through her own ingenuity. This shows us that Betty is capable not only of tool use, but also of tool-making.

1| A small pail of food is placed out of beak's reach at the bottom a transparent tube. Betty takes a piece of straight wire and bends it to make a simple hook.

2| She inserts the wire into the tube and hooks it under the arched handle of the pail.

3| Betty lifts the pail out of the tube with her wire hook.

4| Once Betty has lifted the pail, she can collect her food reward.

2| To reach the food, Betty must insert the stick into one end of the tube and push the food out of the other end.

3| Here, Betty has successfully solved the stick and tube puzzle, and she can be seen claiming her food reward.

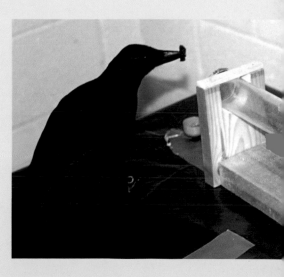

SEA OTTERS HAMMER THEIR WAY TO A MEAL

NATURAL HABITAT
Coasts of the northern and eastern North Pacific Ocean

SEA OTTERS ARE CLEVER CREATURES, WITH AN INGENIOUS ABILITY TO USE ROCKS, ABUNDANT IN THEIR SURROUNDINGS, AS TOOLS, ALLOWING THEM ACCESS TO MANY TYPES OF FOOD THAT WOULD OTHERWISE BE INACCESSIBLE.

Sea otters are the smallest of the marine mammals, and spend a good part of each day foraging for a variety of shellfish and other prey available in the waters of the Pacific Ocean, off the west coast of the United States, from the state of California to Alaska. They use rocks placed on their stomachs to break open the shellfish, allowing them to utilize food resources that other species in the same eco-niche cannot access. Consequently, the sea otters' tool-using prowess can contribute towards their overall health and survival.

Dual purpose

No-one knows exactly when sea otters first began using tools to help them get at their dinner, but they now have a great system for using rocks efficiently, and in two different ways. First, otters use the rock surface as an anvil against which they crush or crack open shellfish, but they also use rocks as hammers, to dislodge animals from rocky areas underwater. Their paws are also great for digging, which is useful since some molluscs, such as clams, can burrow up to 10 cm (4 in) into the sand when trying to escape.

The process of tool use by sea otters is a remarkable sequence of events. After diving to the ocean floor, the otter grabs its prey and searches for a rock, which it tucks under its arm. Otters' armpits are fleshy and pouchlike, and serve as a perfect toolbox. With the prey held in its mouth, the otter swims to the surface and floats on its back. It puts the rock on its stomach and smashes the mollusc or sea urchin against the rock. The hard surface of the rock serves as an anvil, and its webbed paws and claws allow it to get a tight grip as it pounds open its catch.

Otters have an extremely high metabolism, and must eat the equivalent of 30 per cent of their body weight each day to maintain the proper internal temperature. Because the food that sea otters find is very high in protein, they receive a good source of nutrition with each shellfish they obtain.

ECO-WARRIOR OTTER

Some animals that otters feed on, such as sea urchins, can be very destructive to the kelp forests in the ocean. The kelp beds serve as food for snails and fish, and also provide protective cover for many species. Otters eating sea urchins allow the kelp beds, and the species that rely on them, to flourish. This helps to maintain balance in the ecosystem shared by all the species within the otters' domain.

A CRACKING SNACK

Sea otters are the only marine mammals to use rocks as tools.

A California sea otter floats on its back and balances a rock on its stomach, ready to smash open crabs, clams and other shellfish when the opportunity arises.

See also
The wild chimp's toolkit,
 page 34
Wild capuchins adapt tool
 use to suit their
 environment, *page 48*

SUBMARINE SENSATIONS

Sea otters are hefty compared with the other members of the weasel family, weighing anywhere between 14–45 kg (30–100 lb). They can walk on land, but are also capable of living exclusively in the ocean.

Otters are built to dive, with webbed front paws that have semi-retractable claws, and powerful back feet, also webbed, that function more like flippers. The flexibility and streamlined features of their bodies allow them to manoeuvre quickly in the water, making them swift divers. Sea otters are also equipped with lungs that have twice the capacity of a land mammal of comparable size. This adaptation allows them to reach depths of 18–55 m (60–180 ft) to forage for 40 different marine invertebrates, including sea urchins, abalone, crabs, clams, crustaceans and some species of fish.

1 | The otter dives underwater searching for an appetizing morsel. When it spots a clam or similar mollusc it grabs it with its paws then holds it in its mouth.

2 | Next it searches for a rock that it tucks into its armpit pouch.

3 | The otter swims to the surface of the water and promptly floats on its back, placing the rock on its stomach.

4 | It takes the clam in its paws and smashes it against the rock on its stomach. It crushes the shell with repeated blows, and finally retrieves a fat, tasty clam.

SEA SPONGES PROVIDE PADDED PROTECTION

THE FIRST EXAMPLE OF TOOL USE BY MARINE MAMMALS WAS OBSERVED AMONG THE WILD FEMALE DOLPHINS OF WESTERN AUSTRALIA, WHO USE SPONGES TO PROTECT THEIR BEAKS WHILE FORAGING, A BEHAVIOUR THAT IS PASSED ON TO THEIR DAUGHTERS THROUGH OBSERVATIONAL LEARNING.

NATURAL HABITAT
Worldwide, mostly in the shallower seas of the continental shelves

Perhaps the last animal species one might expect to see using tools would be members of the marine mammal family. Yet marine biologists studying wild bottlenose dolphins off the coast of Western Australia, at a place called Shark Bay, observed just that.

Dolphins living near Monkey Mia Beach in Shark Bay began coming close to shore years ago, and slowly they became less and less afraid of the people swimming there. When visitors began feeding them, the dolphins chose to spend more and more time in the shallow waters just off the beach. Researchers with interests in dolphin behaviour soon began a long-term study of the animals at Monkey Mia, following them farther out to sea to observe their natural social interactions and activities. The dolphins were accustomed to being around humans, so they paid little attention to the scientists following them around. This arrangement resulted in a bonanza of observations never seen before in a wild population of dolphins – or any dolphins, anywhere, for that matter.

A curious behaviour

Dolphins have a beak-projection on their heads, known as the "rostrum", that extends about 7.5 cm (3 in) from their body, with a lateral crease along the jawline. Their eyes are located on the sides of their head, very close to the corners of the mouth. The researchers were puzzled for some time after first seeing an adult female dolphin with a sea sponge around her rostrum. They did not know if this behaviour was unintentional; they speculated that perhaps the animal was feeding near a bed of sponges, and accidentally got the sponge stuck. However, more sightings of the same behaviour emerged, and the scientists continued scratching their heads, trying to come up with a reason for this curious activity. Perhaps it was some kind of dolphin tag, or was it possible that "sponging", as they came to call it, had some real function?

THE NATURAL NOSE GUARD

Female offspring learn how to use sponges as protective tools by observing their mothers' actions.

1| Before foraging, a female dolphin searches among a bed of sponges.

2| She selects a cone-shaped sponge to fit around her beak, as her daughter watches.

3| Next, she searches among the coral where small fish and other prey may be hiding, all the while using a type of natural nose guard.

BEHAVIOURAL PATTERNS
The technique of "sponging" seems to be culturally transmitted within a pod through social learning from mother to daughter.

See also
The wild chimp's toolkit,
    page 34
Monkey see, monkey do?,
    page 90

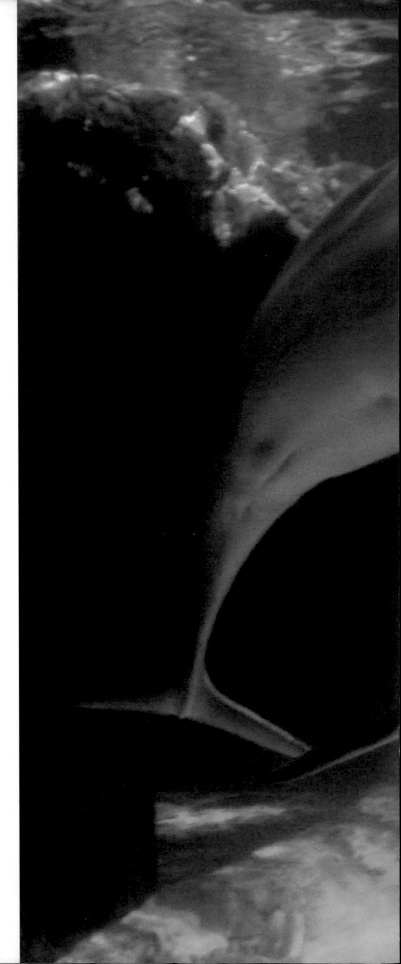

Once the researchers watched the dolphins underwater, the usefulness of the sponges became clear. In addition to the various strategies they use for capturing fish, dolphins also feed in the bottom sediment of the ocean. They can be vigorous diggers, working into the bottom as far down as the pectoral fins on their backs. The prizes they seek are small crustaceans and bottom-dwelling fish. While digging in the ocean floor or amid the coral reefs, a dolphin could unexpectedly meet with a stonefish or other stinging species, or cut its rostrum on the sharp coral. The sponge appeared to act as an insulator of sorts, and helped keep the dolphin's beak from direct contact with dangerous creatures and harmful objects on the sea bottom.

Family affair

Of the numerous sightings of sponging by the dolphins in Shark Bay, all have been of adult females or young dolphins, usually female offspring. It may be the case that male dolphins are too preoccupied with forming coalitions with other males, protecting their females or simply foraging for their own food to bother with sponging.

The social lives of male and female dolphins are quite different. Females and their calves form extremely important bonds that endure for years. Their offspring nurse for three and a half years or more, despite beginning to catch their own fish when they are only a few months old. Even when the calves are no longer nursing, they remain with their mothers for three to six more years. Interestingly, sponging techniques appear to be acquired by young dolphins from their mothers, and thus are culturally transmitted within a pod through social learning. Researchers were intrigued by the pattern of sponging within the dolphin group, and analyzed DNA samples from 13 of 15 animals, including 12 females and a single male, that had been observed to sponge regularly. They hoped to determine if sponging had a genetic basis, or if it was transmitted socially. They compared their results with 172 DNA samples from dolphins that did not sponge. The data revealed that most of the spongers were related maternally, and shared the same DNA that could only have been passed on through the females. Indeed, all of the spongers were closely related, suggesting they were the descendants of a single female. However, the inheritance pattern of the dolphins that sponged did not support any kind of genetic predisposition towards the behaviour. Instead, the results provided support for the idea that sponging was learned by each new generation of the dolphins' offspring, who, in turn, taught their calves to sponge. While no directed teaching of the skills necessary for sponging have been documented, observational learning alone would be adequate for learning to sponge.

SPEEDY SWIMMERS

Dolphins use their protruding rostrum for foraging along the ocean floor. Their eyes are located just behind the rostrum, and they breathe through a blowhole on top of their heads. The dorsal fin on the midline of the back serves as a keel for maintaining their balance as they move rapidly through the water, or jump high into the air. The fluke and pectoral fins are used to propel the dolphin through the water at very high speeds.

NAKED MOLE-RATS PROTECT THEIR ASSETS

NATURAL HABITAT
The grasslands of
East Africa

NAKED MOLE-RATS USE THEIR HUGE FRONT TEETH TO EXCAVATE THEIR BURROWS, AND HAVE LEARNED THAT A SMALL TOOL CAN PROTECT THEIR MOUTHS AND THROATS BY PREVENTING THE INGESTION OF SOIL AND OTHER MATERIALS.

One of the most unlikely candidates for tool use in the animal kingdom is the naked mole-rat. These creatures are native to the drier areas of the tropical grasslands of East Africa, in particular Kenya, Somalia and Ethiopia. They are small, hairless pink or yellowish rodents, about 8–10 cm (3⅛–4 in) in length and weighing approximately 30–35 g (1–1⅕ oz). Their diet consists mainly of bulbs and tuberous roots of plants.

Mole-rats live in underground colonies in a highly organized social structure, with only one breeding female – the "queen" – and one to three breeding males among groups of 12 to 100 individuals. Other members of the colony are classified into two categories. The "workers" spend their days digging tunnels in search of food, while the "soldiers" guard the colony against predators. This type of social structure, while common in insects such as ants or bees, is very rare among mammals.

A captive audience

Since their colonies are located in extensive subterranean burrows, it is not surprising that the mole-rats' tool-using abilities were not discovered until scientists began studying them in captivity, where they could watch them burrow while housed in a series of transparent tubes, with separate nesting boxes and toilet boxes. In their natural habitat, the soil around them is dense and hard, and the mole-rat workers use their large front teeth to dig through the tough dirt. To keep dirt and other objects out of their mouths, they may place a piece of wood shaving or the husk of a tuber behind their front teeth and in front of the lips before they begin digging, as observed first in studies in captivity.

PROTECTIVE TOOL IN ACTION

Using its large front teeth to dig through the hard soil in its natural habitat, the naked mole-rat hopes to find a large tuber or bulb that will feed the colony for weeks.

Naked mole-rats are pale and wrinkly, and are virtually blind, with only small slits for eyes. They are proficient diggers, and tunnels constructed by worker rats can stretch up to 5 km (3 miles), if placed end to end.

1| To protect its mouth and throat from foreign objects or dirt before digging, a worker mole-rat places a wood shaving or husk from a bulb behind its all-important teeth.

2| As the mole-rat excavates another tunnel in search of food for the colony, the wood shaving or husk serves as a tool, helping to prevent choking or ingestion of soil and other objects it encounters while digging.

3| When the mole-rat finds a suitable food source, it uses its large front incisors to tear off pieces that are returned to the colony to feed the breeding animals and the numerous litters of pups born once a year.

WHAT BIG TEETH YOU HAVE

A very young mole-rat pup, with eyes not yet open, holds a piece of food in its paws, ready to start chomping with its huge front teeth.

See also

Sea sponges provide padded protection, *page 26*

SURVIVING A LIFE UNDERGROUND

Naked mole-rats have several specializations that allow them to survive in their harsh underground environments.

• They have no pain receptors in their skin and, highly unusual for a rodent, also have an almost cold-blooded metabolism. Because they can't regulate their temperature, it is crucial that they live in an environment where the temperature is highly stable.

• Their lungs are very small, and their respiration rate is very, very low, which is especially important with the limited availability of air inside the tunnels where they live. However, their blood has a very high affinity for oxygen, which greatly increases the efficiency with which they are able to utilize a limited air supply.

• Adding to their ability to survive in a subterranean environment, mole-rats also have a low metabolic rate. This feature may contribute to what is an unusually long life for rodents: up to 20 years for breeding mole-rats, but only three years for workers, while captive mole-rats have lived as long as 28 years.

• Scientists have speculated that their low metabolism may only contribute small amounts of potentially damaging oxidative processes to their organs or bodies over a longer period of time, and they also spend a great amount of time sleeping.

In the experiments where the first observations of tool use by naked mole-rats were made, scientists created artificial barriers within transparent tunnels to observe how the rats might go about digging passages in the wild. To observe this behaviour, the scientists added barriers made of cork, clay, styrofoam, baked sandstone brick – created by the investigators from sand, water and non-toxic glue – and, inadvertently, plastic. Early in the study, the scientists observed some individuals pick up a wood shaving or tuber husk and place it in front of their teeth before they began to gnaw on the plastic tubes or seams of the nest boxes. In between bouts of gnawing, the mole-rats would scratch at the plastic with their claws. The animals would reposition the shaving or husk, or replace it when necessary. They only used such teeth barriers when gnawing on plastic or brick, and never when the barrier was composed of cork, styrofoam or clay, all of which readily broke apart when the rats began digging against them. Larger individuals were observed to use husks or shavings more often than smaller ones, perhaps indicating differences in age and/or experience, as no rats younger than three years of age were ever seen to use them.

WHY ELEPHANTS USE SWITCHES

NATURAL HABITAT
Africa, India, Sri Lanka, Southeast Asia, Malaysia, Indonesia and southern China

ALTHOUGH ELEPHANTS HAVE BEEN OBSERVED USING TOOLS SINCE THE TIME WHEN CHARLES DARWIN WAS WRITING, MORE RECENT STUDIES OF CAPTIVE ASIAN AND AFRICAN ELEPHANTS AND WILD AFRICAN ELEPHANTS HAVE REVEALED THAT THEY MODIFY BRANCHES AND TWIGS TO MAKE SWITCHES THAT KEEP THE PESKY FLIES AT BAY.

Elephants are part of the wonderful landscape of the savannahs, forests and even the deserts of Africa, as well as forests in some parts of Asia. This magnificent species takes advantage of a number of its own body parts to keep its skin in good condition, such as using its trunk to spray water to clean itself or splashing mud to protect its skin from the sun. Elephants are also known to use their trunks as built-in tools for carrying things, including, under emergency conditions, their own young infants. Their massive ears are useful as fans under the intense sun, and also figure prominently in some types of communicative displays. But elephants are also able to use objects from their environment as tools, and have been observed using branches or clumps of vegetation to keep flies and other insects at bay.

Initial observations

Recent observations of captive animals in a park in Nepal revealed that a group of Asian elephants that were used to taking tourists for rides were grabbing branches or vegetation with their trunks and using them as switches, either to stop flies landing on them or displace those that already had. If branches were offered to them in their home quarters they also used them as switches. This raised two intriguing questions as to the function of the branch tools that were used. Did they help keep the elephants cool, or did they serve to dislodge or keep flies away?

SOCIAL BEASTS
The elephant's massive stature and instantly recognizable trunk are familiar to people around the world. Elephant society is matriarchal; the groups are highly social and ruled by an older, dominant female.

NATURAL SUNBLOCK
Their flexible trunks mean elephants are quite adept at covering themselves in mud. This practice is actually of great value to the elephants, and provides protection for their skin from the blazing midday sun.

LOCALLY MADE
Elephants mainly make branch switches from bushes that are common in their environment, although different species of bushes are used at different locales.

Detailed studies

To answer these questions, scientists conducted a systematic study of elephant behaviour under different conditions. Switch use was defined as the elephant grabbing vegetation or a branch by its trunk, and flailing it against some body part. These included the front and sides of the head, shoulders, legs, chest and the underside of the stomach. When using a leafy branch, the elephants held the switch by the woody end, and used the foliage portion against the body.

In the first experiment, a group of captive elephants was provided with switches that were similar to those that were broken off from trees and bushes by the elephants themselves. An animal might modify a branch by holding it in the mouth or front foot and using the trunk to pull off side branches, or use the very tip of the trunk to break off parts of the stem in order to shorten the branch. Once the elephants had been given branches, the scientists recorded the amount of switching movements that occurred at different times of the day, such as early dawn, when there were few flies present, and throughout the day at 8 am, 11 am, 3 pm and 6 pm, when fly counts differed in their density around the animals. Compared to the number of switching movements made at dawn, which averaged around 30 switches every 10 minutes, the elephants increased the amount of switching movements to 150–186 switches every 10 minutes. This dramatic change in the amount the switches were used was also compared to daytime temperature changes and the elephants' feeding times, and suggested that the behaviours were unrelated and not a function of boredom due to confinement. The elephants still used their switches just as frequently, even when feeding.

A second experiment was conducted among another group of elephants at around 11 am, when fly infestation was at its highest. Tail switching was observed when the animals had no branches to use as tools, and again after branches were given to them. A count of the flies present once the elephants had switching tools showed a 43 per cent reduction in their numbers. All of this evidence suggests that the elephants use the switches to repel and dislodge biting flies.

THE WILD CHIMP'S TOOLKIT

NATURAL HABITAT
Western and Central Africa

MORE THAN ANY OTHER ANIMAL SPECIES, EXCEPT HUMANS, WILD CHIMPANZEES USE THE MOST EXTENSIVE VARIETY OF TOOLS, AND NEW TYPES CONTINUE TO BE DISCOVERED EVERY YEAR.

Perhaps the most incredible scientific discovery of the past century was made by a former secretary, alone in the jungle, watching wild chimpanzees. Dr Jane Goodall reported in the early 1960s on her observations of chimps modifying sticks and twigs by stripping off the leaves, then using the tools to poke into termite mounds. The termite residents "attacked" the invading stick tool by grabbing onto it with their pincers. With a seemingly effortless swish, the chimp would pull the stick out and slide it across its mouth like a termite shish kebab. Suddenly, humankind's "special" place in the animal kingdom was challenged. Our species had been referred to at that time as "man the tool user", and "man the tool-maker", yet in one fell swoop, Goodall's remarkable findings upset these definitions of humanity.

A vast toolkit
Following Goodall's original report of chimpanzee tool use at her study site, Gombe Stream Reserve in Tanzania, other sites in different African countries began to report the use of different kinds of tools. Across the different "toolkits" (various types of tools used)

PESTLE POUNDING

The chimpanzees of Bossou in Guinea, West Africa use a clever technique called "pestle pounding" for processing the centre of the crown of an oil-palm tree, before eating the pith.

1 | A chimpanzee uses its hands and feet to prise apart the uppermost branches of an oil palm tree.

2 | The ape detaches a palm frond to use it as a pestle to pound and soften the pith at the centre of the crown.

3 | Finally, the chimpanzee scoops up the pounded pith in its hand.

See also
Use of spears by wild
  chimpanzees, *page 38*

EXCAVATIONS
The chimpanzees at
Bossou have been observed
pounding and excavating
the centre of oil-palm trees
in order to reach the tasty
palm heart.

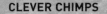

and tool use documented among the chimpanzee communities
that have been studied long-term, some 35-plus tool types have
been identified so far. For example, studies have revealed that
wild chimps will remove all the fronds from fern stalks in order to
collect algae from water surfaces, and will also use a number of
chewed-up leaves to from a "wadge" that acts as a sponge
when reaching water in the holes of trees. Even more exciting are
the most recent findings that chimpanzees modify and use simple
spears to catch bushbabies from deep within the tree nests made
by the tiny animals.

More recently, chimpanzees have also been observed in several
different locations in Africa to use stone tools as hammers and
anvils for cracking open oil-palm nuts. Oil-palm nuts have a very
hard outer shell, and without the use of stone tools, the chimps
would not be able to access this rich food source. Yet not all
chimpanzees in Africa show stone-tool use, even when there
are oil palms in their habitat.

Comparisons of all the chimp sites across Central and
Western Africa and their respective toolkits have revealed that
chimpanzees may have a rudimentary culture, since they use a
range of different tools at each site, but not all tools are used
at all sites. This phenomenon suggests that chimpanzee groups
acquire expertise with the types of tools used within their specific
community through observational learning from their mothers.

Watch with mother

Detailed observations of wild chimps have also revealed sex
differences in their use of tools. Female chimpanzees are the
most consistent tool-makers and tool users, by far. Adult male
chimpanzees do not use tools very often, and consequently
do not depend on food obtained by using tools. Female
chimpanzees, however, depend upon a wide variety of tools
that are made and used creatively for acquiring a significant
portion of their diet.

HUNTING PARTY

Common chimpanzees are omnivores and eat anything that is non-toxic, including birds, eggs, flowers, fruit, foliage and small mammals, as well as, if the opportunity arises, some species of monkey whose habitats overlap with the chimps' territory. The chimps are known as "opportunistic omnivores", which means that no planning is involved in their hunting of live prey. Chimpanzee hunting parties are usually all male, and if an unaware monkey wanders into their area, they may take advantage of the opportunity at hand. They quietly gather together to hunt cooperatively by encircling the fated animal, driving it towards the other chimps in the party until it is caught. The meat is shared among the males, with the victorious chimpanzee who caught the monkey doling out the spoils, including tolerating begging by females.

Both young males and female chimps observe their mothers using tools over a period of three to seven years, and practise copying her techniques, though often without success in their early attempts. Years later, they will become proficient.

No tools for the cousins

Surprisingly, the bonobo or pygmy chimpanzee (*Pan paniscus*), a different species from the common chimpanzee (*Pan troglodytes*), has not been seen to use tools in its natural habitat, in dramatic contrast to the extensive tool repertoire of its common cousins. The two chimpanzee species share the same taxonomic genus (*Pan*), but are separate and distinct species. Currently, the highly endangered bonobo is only found in the Democratic Republic of Congo, where they live in protected reserves.

The habitat of bonobos may offer some clues to help explain why tool use is not seen in the wild. First, bonobos are more arboreal than their terrestrial cousins, and may find all they need to survive among the trees and bushes in their territory. Consequently, it may not be necessary for them to exploit inaccessible food sources on the ground through the use of tools. Bonobos have also not been observed to hunt for meat, another indication that their diet, derived from fruits, eggs and foliage, is sufficient without additional protein sources such as termites, ants or meat from small mammals or monkeys sought out by common chimpanzees.

CLOSE RELATIONS

Chimpanzees are our closest relations in the animal world. Like us, they possess opposable thumbs which facilitate their use of tools.

USE OF SPEARS BY WILD CHIMPANZEES

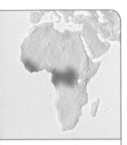

NATURAL HABITAT
Western and Central Africa

NEW OBSERVATIONS OF WILD CHIMPANZEES IN SENEGAL, WEST AFRICA, REVEAL THAT THEY MANUFACTURE AND USE SPEARS TO HUNT SMALL PRIMATES – THE FIRST-EVER DOCUMENTED SIGHTINGS OF WEAPON USE BY APES.

Nothing could have prepared a small group of anthropologists studying chimpanzees in Senegal for what they saw happening at their study site. They could hardly believe their eyes, but chimpanzees were fashioning functional spears from sticks and using them as weapons. The chimps' target was the bushbaby, a small primate from the family known as prosimians, meaning "before monkeys". Bushbabies are abundant throughout Africa, and are popular as pets because of their small size and cuddly features. However, the chimpanzees at the Fongoli site in southwestern Senegal, who have been studied for seven years by researchers, were not looking for cute companions. They wanted dinner.

Talented toolmakers

Hunting with spears begins when a female chimpanzee selects a stick that is the appropriate size and diameter. Next, she gnaws on the end to fashion a spearlike tip. Once she has identified a potential site in the hole of a tree – where bushbabies often nest – the hunter plunges the spear into the hole. If no bushbaby is captured when she withdraws the tool, she smells the end of the spear, probably for blood. If the chimp is successful, a bushbaby has been caught and is eaten on the spot. The chimpanzees do not have to actually see the prey, but instead have a mental representation that suggests the likelihood of lunch being found inside the tree. Without the use of spears, any disturbance of a bushbaby nest sends the little critters scattering quickly beyond the reach of the chimpanzee. The spear acts as an extension of the chimp's arms and hands, and this represents part of the definition of a tool.

The use of spears by the Fongoli chimps is a type of extractive foraging that is similar to ant dipping or termite fishing observed in chimpanzees in Tanzania, where tools are necessary to "extract" the food resources from an otherwise inaccessible spot.

Female innovation

It is probably not a coincidence that only females or very young chimpanzees have been observed using tools. In other areas across Africa where studies of chimps have been ongoing for

HUNTING AID

Hunting with spears requires a degree of planning by the female chimpanzee.

1| First she selects a strong stick, then uses her teeth to whittle the end of it. Next, she looks for a likely hideaway for a sleeping bushbaby, that is active only at night.

2| Grasping the other end of the stick, the chimpanzee hunter forces the stick down into the cavity of the tree, hoping to skewer her dinner.

3| The stick becomes an extension of the chimp's arm and allows it to secure an out-of-reach meal.

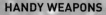

HANDY WEAPONS
Although spear use is very rare and has only recently been observed, chimpanzees quite readily pick up sticks and large stones to use as weapons.

See also
The wild chimp's toolkit,
    page 34
Monkey see, monkey do?,
    page 90

decades, it is females who are the most avid tool-makers and tool users. These observations have suggested to paleontologists and archeologists that the innovative tool users among the earliest humans were almost certainly females. This hypothesis also assumes that it was early hominid (human-like) females who manufactured the tools as well.

The scientists studying the chimps have speculated that the females and younger animals may have had difficulty competing with the adult males for food. Therefore, tool use emerged among them as an innovation for obtaining additional protein.

Fongoli chimps stand out from the crowd

The Fongoli chimpanzee group has also distinguished itself by other novel behaviours that have not been seen in any other chimp communities in Africa. The study team has reported that the chimpanzee group uses cool caves in their territory to escape the African sun on the savannah, where they also bring food for an impromptu picnic. Even more remarkable is that the group is not afraid of water, and has been seen splashing and wading into streams and pools. Chimpanzees in other parts of Western and Central Africa avoid water, and, in fact, it had been assumed previously that they had a natural fear of water. With little body fat, a chimpanzee would sink quickly in water that was too deep, and consequently, avoiding water appeared to be a highly adaptive behaviour. But remarkably, the savannah-living chimpanzees of Senegal have defied many of the old assumptions about chimpanzee behaviour reported by field researchers at other sites. Here, females make and use weapons for hunting smaller prey; the group makes use of caves to escape the burning sun; and they also appear to enjoy play in water, in dramatic contrast to chimpanzees studied elsewhere.

SUMATRANS GET TO GRIPS WITH TOOLS

NATURAL HABITAT
Borneo and Sumatra

SCIENTISTS STUDYING ORANG-UTANS IN BORNEO WIDENED THEIR INVESTIGATIONS TO SUMATRA AND DISCOVERED THAT NOT ALL ORANG-UTANS ARE THE SAME, AND LACK OF TOOL USE IN ONE GROUP DOES NOT MEAN THAT ANOTHER GROUP HAS NOT MASTERED THE ART.

Object manipulation and tool use in orang-utans was shown to be a prominent skill among those kept in captivity, and similar skills soon emerged in orphaned orang-utans that were brought to rehabilitation centres in Borneo. However, these observations revealed a great puzzle, since the long-term study of wild orang-utans had not documented any type of tool use, and for the most part, the animals were fairly solitary. Food resources were scarce and widely distributed, so foraging alone was critical for survival in order to not compete with other orang-utans. However, very recent observations of wild orang-utans on the nearby island of Sumatra paint an entirely different picture of this species.

Vive la difference

While there are differences in the types of orang-utans that live on the two islands of Sumatra and Borneo, their overall physical attributes are highly similar. It had therefore been assumed for decades that the orang-utans on Sumatra acted in much the same way as those on Borneo. However, that was not the case, and field researchers who withstood a two-hour hike to and from the site, waist-high in water and surrounded by leeches and flocks of mosquitoes, were privileged to discover startling differences, and saw orang-utans in a whole new light.

On Sumatra, in what has been described as a highly productive swamp habitat, orang-utans not only use tools, they use several types of tools for extractive foraging. In order to get to the high-protein seeds and pulp located deep in the centre of the neesia fruit, a tool is necessary. Without stick tools to break into the fruit, the orang-utans would not be able to get to the

ENJOYING THE FRUITS OF THEIR LABOUR
A wild Sumatran orang-utan female uses a stick for a tool, and is relishing the inside pulp and seeds of a neesia fruit. While neesia trees are prevalent in their habitat, the interior of the fruit is covered with tiny, thin, but extremely sharp needle-like hairs, which means that using fingers to pry out the seeds would be very unpleasant.

FOREST VOICES
Orang-utans are highly recognizable by their long, red hair. Adult males are considerably larger than females and have large cheek pads around their faces that help them project vocalizations over long distances in the forest.

See also
The wild chimp's toolkit, *page 34*
Wild capuchins adapt tool use to suit their environment, *page 48*

inside pulp that is rich in fats and protein. Using their hands or teeth would be very painful because the edible seeds in the centre of the hard-shelled fruit are surrounded by tiny, needle-like hairs. Consequently, using a tool allows the orang-utans to exploit a food resource that is not accessible to other animals.

In addition to using sticks, orang-utans were seen using other tool types. For example, they modified sticks very effectively into useful tools for reaching honey or insects deep inside tree holes, but used a different kind of stick when opening fruits.

Unlike in other environments where tool use provides a significant contribution to the animals' daily food, and is truly necessary for their survival, the swamp forest habitat of the Sumatran orang-utans is quite rich. However, using tools allows for greater variety and access to other nutritional food sources. The abundance all around them allows the animals to remain in close proximity to one another, rather than resorting to solitary feeding that had been the only foraging strategy previously seen in orang-utans.

A group tradition

The scientists wondered if orang-utans used tools wherever there were neesia trees, and made observations of orang-utans living across the river from their original site in a smaller swamp. Neesia trees were there, but there was no tool use by this group of orang-utans, so the fruits were not a part of their diet. Yet, just across the river in the larger swamps, orang-utans used tools in abundance. This difference demonstrated that tool use among the orang-utans on one side of the river was a cultural tradition, observed by each new generation of young orang-utans, and incorporated into their behavioural repertoire. The animals on the

COMMUNAL LIVING
It used to be thought that orang-utans live entirely solitary existences. However, in Sumatra groups of up to 100 individuals sometimes congregate in the forest.

other side of the river were unable to cross it, and consequently did not have the opportunity to learn about tool use and tool-making from others in their community.

Another striking difference that the researchers found among the Sumatran orang-utans was the extent of their social interaction. They were highly social and extremely tolerant of one another, such that they would gather in groups as large as 100 animals. A gathering of such a large number of orang-utans was unheard of before the discoveries on Sumatra. There, orang-utans of all ages foraged together, using tools when necessary, and even shared food with one another. None of these behaviours were ever observed among the Bornean orang-utan populations.

These unique findings provided another clue as to why orang-utans under captive conditions were so inventive and could also be highly social. Despite the dedication of the scientists who had been studying the orang-utans on Borneo, their reports on their subjects' social behaviour, foraging strategies and lack of tool use in the orang-utan did not fully describe its great cognitive potential as a social tool user.

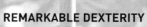

REMARKABLE DEXTERITY
The orang-utan has enormous hands, with each finger the size of a large banana. It is remarkable that despite this huge hand size the orang-utans on Sumatra are proficient tool users.

WILD GORILLAS STUN RESEARCHERS

NATURAL HABITAT
Western Central Africa

IT IS ONLY VERY RECENTLY THAT A LIMITED NUMBER OF OBSERVATIONS OF WILD GORILLAS USING STICKS AS TOOLS HAVE BEEN REPORTED.

Historically, when compared to the other three great apes, gorillas have usually been unfairly relegated to the bottom of the pile when it comes to intelligence. Perhaps this is, in part, because their behaviour is gentler, shyer and less aggressive than that of chimpanzees. Gorillas in the wild do not hunt for meat like chimpanzees, but instead lead a vegetarian lifestyle. Their social structure is also radically different from that of the other apes, with one dominant silverback male overseeing a harem of breeding females and their immature offspring. This arrangement may keep the day-to-day interactions at a manageable level, because all the members know their place within the group, and gorillas have also learned that violations of the social rules will be dealt with swiftly and firmly by the massive silver-backed adult males.

Proving them all wrong

Does the range of behavioural differences between gorillas and other apes mean that the gorillas do not need to be smart? Not necessarily, say field researchers who reported the first observations of tool use by wild gorillas in 2005. Working in the Congo, the scientific team had been observing Western lowland gorillas at the site for over 10 years, and had never seen any tool use before the two sightings, each of which represented a different type of tool use. The researcher was shocked to see an adult female they had nicknamed "Leah" wade into a deep waterhole left behind by elephants. When the water got up to her chest, she reached above for a branch and broke it off. Leah began using the long branch as a walking stick; she tested the depth of the water with it, and continued to monitor the depth of the water using the tool, walking a full 33 m (108 ft) before she returned to the shore.

GORILLAS IN THE WILD
Gorillas are the quiet giants of the forest, but their numbers are depleting across Africa as deforestation and logging destroy acres of rainforest every day.

The second observation was equally spectacular, as "Efi", a female gorilla from a different group, was seen taking the stump from a bush and using it as a seat as she used a long stick to dig for herbs. In Efi's case she was using a tool to stabilize her body so that she could use another tool for another purpose.

The findings were a tremendous surprise for the scientific community, and photographs of the two tool-use events travelled the world via the media. Indeed, relative to the thousands of hours spent over decades of observations of wild gorillas, the two sightings represent a paucity of tool-use activity by wild gorillas. Nonetheless, the documentation of three different functional uses of natural materials as tools by wild gorillas was unprecedented. Of added interest is the fact that the two animals observed to use the tools were both adult females, because sex differences in tool use by wild chimpanzees has been reported on numerous field researches. It is female chimpanzees who are the most consistent and creative tool users, and their traditions are passed on culturally to their offspring through observational learning over three to five years.

GORILLA BATHING BRANCH

Great apes, including gorillas, are poor swimmers and, whenever possible, they avoid entering deep water. And all great apes, except for humans, find walking on two legs for any length of time difficult. Therefore, a branch used as a combined walking stick and probe, testing the depth of water in a pool, is an extremely useful invention. The first observation of a wild gorilla making and using a tool involved this probing walking stick.

1| A female gorilla, Leah, began wading through a deep pool but she soon returned to the shore when the water reached waist height.

2| Leah broke off a long branch of a dead tree and returned to the pool.

See also
The wild chimp's toolkit, *page 34*
Use of spears by wild chimpanzees,
   page 38

3| She then waded 10 m (30 ft) out into the pool,
constantly probing the depth of the water with
her makeshift walking stick.

STUNNING SIGHT

Leah was the first ever wild gorilla observed making
and using a tool, a sight that stunned researchers.

WILD CAPUCHINS ADAPT TOOL USE TO SUIT THEIR ENVIRONMENT

NATURAL HABITAT
Central and middle
South America

WILD CAPUCHIN MONKEYS USE STICK TOOLS IN THE RAINFOREST FOR FINDING INSECTS, BUT MONKEYS LIVING IN DRIER HABITATS HAVE MASTERED THE ART OF USING ROCKS AS HAMMERS FOR CRACKING NUTS.

Capuchin monkeys are about the size of a cat, weighing 1.3–2.3 kg (3–5 lb), and are found throughout Central and South America. Capuchins have long been known to easily learn to use a variety of tools under experimental conditions. There have also been numerous reports from field researchers describing how wild capuchins have used tools to locate food that is not visually available. They have observed monkeys using small wooden limbs to pound on rotten logs to uncover grubs, and using sticks or twigs to access insects or larvae from crevices or holes in trees.

Most field studies were conducted within the tropical rainforest, where seasonal fruits and other foods that are part of the capuchins' diet are plentiful. Many food resources are readily available, so using tools to acquire extra foods is not always necessary in this area. However, very recent discoveries of wild capuchin monkeys in Brazil, living in a harsher, drier and more desolate environment, have revealed the monkeys' use of rocks as stone tools for opening hard nuts. Using tools allows the monkeys to supplement the meagre diet available in this habitat, providing them with a rich source of protein. When primatologists began observations of monkeys living under these conditions, the more specialized tool use displayed by the animals had never been monitored before.

Using stones

The monkeys use surprisingly heavy stones to open palm nuts, which they bring to special rocky flats that they use as anvil sites. The anvil sites have been used so often that shallow "bowls" have been worn into the surface of the rock. The distribution of multiple bowl-like indentations in the area suggests that a significant number of monkeys may be using the areas at the same time.

In terms of techniques with the tools, researchers have seen the monkeys open palm nuts in two different ways. Despite the hefty size of the rocks, some monkeys were observed to stand

CAPUCHIN FEATURES

As is characteristic of all New World monkeys, the nostrils on a capuchin monkey's nose point to the sides. There are a number of species of capuchins that have different-coloured hair around the face, and others that have black hair on their heads that appears to be fashioned as a well-coiffed topknot!

See also
The wild chimp's toolkit, *page 34*
Innovations with tools in captive
capuchins, *page 50*

BIG IS BEST
Rocks of many sizes are plentiful,
but surprisingly the capuchins
typically select extremely heavy
stones in relation to their small
body weight.

upright on both legs while pulling the stone up to their
shoulders, then letting it crash onto the nut, crushing it. The
second technique was more demure, with the monkey either
sitting or standing while holding the rock with both hands and
hitting the nut with several repeated up-and-down motions using
the arm and shoulder muscles.

Given the extensive demonstrations of capuchin tool use in
captivity, it should not be surprising that scientists have finally
discovered significant types of tools used by capuchin monkeys
in the wild. In fact, the monkeys' tool-using abilities were known
by the local people for decades, but it was only recently that
systematic observations confirmed and documented the use of
stone tools by wild capuchins. Also notable was that the entire
population of monkeys that were studied used stone tools, rather
than just a few innovators. Previously, this type of tool use had
only been observed in wild chimpanzees in West Africa, and then
only in females and youngsters.

Another interesting twist that remains to be explained is
where the rocks come from, since they are large and worn
smooth by water. Similar rocks have not been found in the

area where they are used, so future studies will investigate
whether the monkeys actually transport the stones from
another area, perhaps some distance away.

Sticks and stones

At other sites in Brazil, additional types of tool use have also
been documented. In a dry forest area, capuchins used twigs
and sticks, which they often modified first, as probes for catching
insects. These capuchins also used stone tools in different shapes
and sizes for digging and dislodging their intended prey. The
researchers argue that because they live in such a harsh climate
the capuchins must compensate for limited food availability by
using tools for "extractive foraging", that is, for finding food that
is not readily visible and/or accessible, but requires the use of a
tool to obtain it.

INNOVATIONS WITH TOOLS IN CAPTIVE CAPUCHINS

NOT ONLY HAVE CAPTIVE CAPUCHIN MONKEYS BEEN SHOWN TO USE TOOLS APPROPRIATELY AND FUNCTIONALLY IN NUMEROUS EXPERIMENTS, THEY ARE ALSO INVENTIVE WITH OTHER MATERIALS.

NATURAL HABITAT
Central and middle
South America

Scientists have long been fascinated by captive capuchin monkeys, who have distinguished themselves as exquisite tool users, demonstrating intriguing innovations, using different types of material, and solving challenging problems that required the use of novel tools. Capuchins were chosen as subjects for these types of experimental studies because of their remarkable manipulative abilities, which had been observed in captive settings for centuries. Native folklore in their natural habitats in Central and South America, and anecdotal observations, though formerly quite limited in the wild, had contributed to the capuchin's nickname of "the poor man's chimpanzee". Certainly, the range and extent of their tool-using capacities studied in zoos and laboratories supported this notion. Indeed, not only did captive capuchin monkeys use tools, they also manufactured them when necessary.

Its all in the hands

The great dexterity of the primate hand, with its five fingers and range of motions with opposable thumbs – all modifications that occurred in the primate order over a long period of adaptive change – allows for the variety and precision with which captive capuchins use tools. In addition, capuchins are unique among the New World species (Central and

THE RAW MATERIALS
The capuchin's precision grip allows it to make and use a wide variety of tools.

South American primate species) because they can move each finger independently, allowing for their demonstrated flexibility in manipulating objects. These changes allow the monkeys to use what is known as a "precision grip" that permits fine motor movements between the thumb and forefinger.

Probe experiment

Under many testing situations, capuchins readily use sticks or other objects as extensions of their arms and bodies, in order to obtain food that is otherwise beyond their reach. Under semi-naturalistic free-ranging conditions in a privately owned zoological garden located in a tropical climate, capuchin monkeys were free to forage for insects, fruit, foliage and invertebrates within their exhibit. Researchers were interested in whether the monkeys could take advantage of the natural materials all around them for use in a specially designed artificial feeding station that required them to use tools as probes. Some of the animals had previous experience with tool use in an experimental study, but had never been tested under more naturalistic conditions. In this case, the researchers were interested to see whether the monkeys would adapt to the more rigorous demands of the task, where access to tools was dependent upon their own manufacturing abilities and recognition of the tool requirements needed for success.

Holes 6.4 cm (2½ in) deep and 0.8 or 1.2 cm (⁵⁄₁₆ or ½ in) in diameter were drilled into a large tree stump and baited with honey. In order to reach the honey the capuchins needed a

GETTING TO GRIPS
Because of their opposable thumbs, capuchin monkeys are capable of a precision grip, similar to that of humans.

functional tool that was at least 6.4 cm (2½ in) long, was sturdy enough to serve as a dipping tool and slim enough to fit into the available holes. Not surprisingly, the monkeys with prior experience at other tool tasks used tools on 69 to 97 per cent of the available test trials. They rarely tried to use inappropriate tools, and thus their tool choice appeared to be non-random. Potential tools were bitten and torn twigs and branches from bushes, trees or nearby vines, which were first broken into smaller sections before leaves and stems were removed. They also removed broken or limp ends and, when necessary, split sticks into more usable diameters. These behaviours likely replicate the types of modifications that wild capuchins would undertake before using tools for similar extractive foraging. These modifications have suggested to some scientists that capuchin monkeys have the requisite cognitive capacity to have a representation – some type of mental template – of the necessary features of the tool they need.

In another study, captive capuchins were provided with bones and rocks, and successfully modified the bones into effective probing tools by using the rocks as stone tools.

Cause and effect

Another creative and unique task was designed by Italian researchers who were interested in determining if capuchins understood something about the causal relationship between the use of a tool, the functional requirements of a tool and the effects of their own actions on the tool. Though numerous demonstrations of tool use by captive capuchins had been shown, the extent to which the animals truly understood how and why they should use certain tools in particular ways was difficult to determine. In a series of experiments, a horizontal, transparent plastic tube was mounted on a raised bracket. A peanut was placed in the centre of the tube by the experimenter while out of sight of the monkey subject. A wooden dowel that could be used appropriately as a tool was placed by the tube and the monkey allowed access. The monkey could insert the tool from either end of the tube, but had to push the peanut away from itself in order for it to fall out the other end. This type of movement, pushing food away to obtain it, was highly counter-intuitive to the monkeys. However, three of the four subjects tested were able to use the tool spontaneously.

More complex test conditions were then introduced to the three successful monkeys, and required them to combine sticks or modify tools in order for them to work effectively. For example, several sticks that were each too short to reach the peanut were provided, and the monkeys needed to put at least two shorter sticks into the tube in order to push the food out. In another condition, a crosspiece had been inserted through

CAPABLE CAPUCHINS
Capuchins are highly social monkeys from Central and South America. They are unique among monkeys because they regularly make and use tools.

DECISIONS, DECISIONS

The reward for the trap-tube task can be placed on either side of the hole leading to the trap, so the monkey must decide on the best side for inserting the tool.

Failure

1| If the monkey inserts the tool so that it reaches the peanut first...

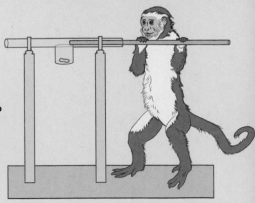

2| ...and then continues to push, the peanut drops into the tube and is lost.

the dowel at each end, so the animals had to modify the tool by removing one of the crosspieces so that the dowel could be inserted in the tube. Under other conditions the monkeys were given a dowel that was too thick, as well as a sheaf of small reeds that would fit, and in another test condition, a stick that was too large in diameter for the tube had to be modified to make a thinner tool. All three capuchins were successful with all conditions in a matter of a few minutes, although each used different techniques for solving the problems. Although the animals proved themselves to be skillful tool users and tool-makers, they often discarded good tools that they used for a short period in favour of less suitable ones, suggesting that they did not have a clear goal in mind for task solution. For instance, they did not modify tools prior to initiating a new problem, but rather, only after they initially failed at using a new tool. This meant that the modifications they made were the result of trial-and-error failures to which they responded by making a change in the tool until they succeeded with it.

The trap-tube task

To explore whether the capuchins could understand the true causal relationships inherent in a complex tool task, the scientists next created a new tube with a hole in the centre that had a small trap attached underneath. Now it was critical that the monkeys pay close attention to the placement of the reward. If the subject inserted the tool into the side closest to the reward and pushed, the peanut would fall into the trap. However, if the monkey inserted and pushed the tool from the other side that was further from the peanut, the tool would slide across the opening to the trap and push the peanut out. Only one capuchin learned to solve the trap-tube task. She always looked into the tube, and adjusted to one side or the other, depending on the peanut's placement. To test if she really understood the problem, the experimenters turned the tube upside down, so that the trap was no longer effective. Under this test condition she used precisely the same strategy as if the trap was still in place, demonstrating that she was using perceptual strategy, rather than really showing she understood the causal features of the task.

Prior to revelations of the different tool-use capabilities of the monkeys in the wild, capuchins were described as "destructive foragers" only, because they gained access to food sources by breaking apart logs, vigorously manipulating potential food sources, and banging logs onto dead wood or other objects, in order to dislodge prey. A similar approach, though ultimately successful with the tube task, indicated that the capuchins did not appear to understand the cause-and-effect relationships inherent in the various tool tasks. Nonetheless, they were successful in using a variety of approaches that likely reflect similar foraging patterns that yield food sources for wild capuchins that are not available to other animals in the same eco-niche.

Success

1| Roberta, the only successful capuchin tested with a non-inverted tube, persisted in inserting the tool so that it reached the trap (and not the food) first...

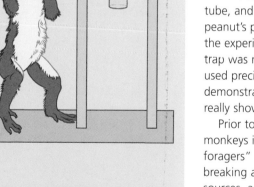

2| ...successfully freeing the peanut.

TOOL USE IN CAPTIVE CHIMPS

NATURAL HABITAT
Western and Central
Africa

CAPTIVE CHIMPANZEES HAVE DEMONSTRATED
THEIR FLEXIBILITY AND CAPACITY FOR USING TOOLS
UNDER A VARIETY OF CHALLENGING CONDITIONS.

Decades before the discovery that chimpanzees make and use
tools in the wild, studies of chimps in captivity documented their
abilities to solve problems by using tools. During World War I, the
German psychologist Wolfgang Köhler began studying captive
chimpanzees in a research station in the Canary Islands. To
understand how they learn to solve problems, he gave them tasks
that required tools for retrieving food. For example, they might
encounter bananas tied to the top of their enclosure, with only
large wooden boxes available nearby. The fact that the bananas
were so high did not deter the chimpanzees, as several of them
cooperated with one another by steadying the boxes while
one climbed up to snatch the prize. It is likely that the same
chimpanzee who grabbed the banana helped other chimps
later by taking a turn at holding the boxes. In another study,
the chimps had access to different sticks that varied in length,
but none of them were long enough to reach food that had
been placed out of reach beyond the edge of the cage. The
chimpanzees quickly learned that one stick could fit inside the
other, making a longer tool that was right for the job. Similarly,
the chimps learned to use a box and a stick together, standing
on the box and using the stick tool to bat at a banana that was
hanging out of reach.

APE LESSONS

Captive chimpanzees and two other apes have
even been taught how to flint-knap, that is, to
use a granite hammer stone struck against a
flint stone to produce a flake with a very sharp
knife edge. In these studies, a special box was
designed that would hold a tasty treat, but
could only be opened by using a knife to cut
a rope holding the door shut. Chimpanzees,
an orang-utan and a bonobo in separate
projects all learned to make a stone flake,
and use it as a knife to obtain a reward.

CLOSE COUSINS

Chimpanzees are our closest genetic relatives, sharing approximately 98.6 per cent of our DNA. They behave in ways remarkably similar to humans, including making and using tools.

CHIMPS CHALLENGED

Köhler devised many different puzzles to test the ingenuity of his group of chimpanzees. He wanted to discover whether they could solve novel problems without going through a long period of trial and error.

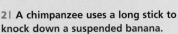

1 | Two chimpanzees steady two stacked boxes, while a third reaches a suspended banana.

2 | A chimpanzee uses a long stick to knock down a suspended banana.

3 | The chimpanzees combine box-stacking and stick use to retrieve a banana suspended high above them.

Precursor to wild studies

In the book that he later wrote about the studies, entitled *The Mentality of Apes*, Köhler proposed that the chimpanzees were able to use what he called "insight learning" to solve the tasks. However, studies completed by other scientists later revealed that the chimpanzees' success was attributable to the various ways they had experimented with using the tools, and therefore that they had used "trial-and-error learning" to work out how to reach the reward. Nevertheless, Köhler's early studies of chimpanzee problem-solving represented a pioneering effort to understand the potential capacity of the chimpanzee's ability to use tools under experimental conditions. No-one could have known at the time how much and how often tool use would figure in the lives of wild chimpanzees, nor the extensive types of tools that now continue to be discovered at field sites all over Africa. These findings reveal that the need and ability for captive chimpanzees to manipulate objects in their environment, particularly as tools for attaining food, is in fact a natural capacity that was first observed outside of their natural habitat.

Sociable studies

Since the unprecedented studies of captive chimpanzee tool use in the early part of the 20th century, a vast array of subsequent studies have been conducted by scientists interested in chimpanzee learning abilities and overall intelligence. While comparisons to human skills are inevitable, given how close chimpanzees are to humans genetically, the chimpanzee is worthy of close study in its own right. Its complex social structure in the wild, extensive use of materials for extractive foraging – locating and obtaining food resources that are not visually available or that require tools for extraction – and exquisite abilities for object manipulation, among many other remarkable skills that are unique to chimpanzees, reveal a species with its own incredible

GO FISH

Chimpanzees use tools in more varied ways than any other non-human animal. Here a chimpanzee uses a branch to fish for a piece of food floating out of reach.

evolutionary journey. Studies that have presented chimpanzees with a range of tasks, from simple undertakings such as matching colours or shapes, discriminating between objects, and cooperating with one another to move heavy objects that could not be moved by a single animal, to more recent long-term studies of representational symbol systems, have shown that chimpanzees rise to the occasion, and meet virtually every challenging task, problem or skill level set before them. It is likely that the other great apes, most of which have not had as extensive testing or teaching as chimpanzees, could show highly similar abilities, but it is the sociability of chimpanzees that has allowed us to get a closer look at their capacities.

Cognitive tests
Numerous research teams have taken an interest in more specific problems related to chimpanzee cognitive abilities. One area of particular interest is whether chimpanzees can imitate the precise

sequence of a tool task that features multiple steps. Previous studies comparing chimpanzee and other ape subjects and offspring, where tool use was first shown by a human model, often resulted in the animals completing the tasks, but not in exactly the same way as the human model had demonstrated. The offspring, however, did match the model's methods, which led the observers to conclude that chimpanzees were unable to show true imitation.

However, a later group of researchers designed tool tasks that replicated the types of problems encountered by wild chimpanzees. When they used this approach, their chimp subjects were successful at completing the task using the same methods as their human models, dispelling the idea that chimpanzees were unable to show imitation that was similar to human abilities. To extend this study, the scientists considered whether chimpanzees themselves could serve as models for a new type of tool problem. They gave three captive groups a novel food box, and taught one high-ranking female from each of two groups to use a particular method with the box. In the third group, no chimpanzee was taught ahead of time how to use the box. Thirty of the 32 chimps that had a chimp model the method for them learned how to use the new food box, while the third group, which had no chimp model to demonstrate, never learned how to use it. This study provided a good experimental demonstration of how tool use might be transmitted culturally, through observational learning from expert tool users, nearly always adult females, in a particular chimp community.

Upping the complexity

Still more experiments have attempted to look more closely at strategies used by chimpanzees when confronted with a more complex tool problem. Initially, capuchin monkeys were tested with a special tool apparatus featuring a clear tube open at both ends, with a middle hole connected to a small cup or trap (see page 52). Depending on which side the experimenter used to push a peanut close to the hole, the monkey could start from one side or the other and use a stick tool to push the peanut out of the tube. The tool had to be inserted from the correct side so that it would move across the opening of the trap, touch the peanut on the other side of the hole and push it all the way to the other

opening. If the stick was inserted on the same side of the hole as the peanut it would push the treat into the trap. However, only one monkey was finally able to solve the problem, but also used the same technique when the tube was inverted so that the trap was no longer functional. The experimenters concluded that the monkey was using perceptual cues in order to know which side to use, and therefore did not really understand the relationship between the trap and how to use the tool. What would chimpanzees do if given the same task?

A group of chimpanzees was given the same test by the same scientists. The chimps easily learned to use a stick to push sweets out of a straight tube, so the trap tube was introduced to see what kind of strategies they would use with a more complicated task. Within 150 trials, several of the chimpanzees inserted the tool into the correct side on most tests. These results indicated that, unlike the monkeys, the chimps recognized what would happen to the sweets ahead of time if they put the tool into the wrong side. This meant that the chimpanzees were able to mentally solve the problem first by representing the various features of the task in their minds before they used the tool. This way, they were able to envision the side that would allow them to attain the sweets reward successfully. The scientists concluded that, unlike the capuchins, chimpanzees had the necessary mental capacity to use a cognitive strategy, rather than the visual cues used by their monkey cousins. The ability to use mental representation for problem-solving and other kinds of tasks or skills is a hallmark of human information processing, and is clearly shared with us in some measure by the chimpanzee.

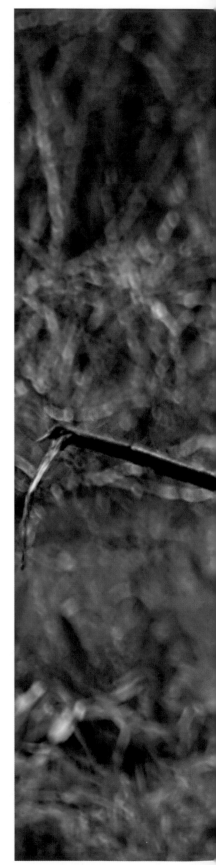

TOOL CRAFTING
Chimpanzees not only use tools, but they modify objects to turn them into tools. Here a chimpanzee prepares a probing stick by stripping off the bark.

See also
Innovations with tools in captive capuchins,
page 50

CHAPTER TWO

COMMUNICATION

Long before humans evolved their capacity for speech, other animal species already had remarkably effective systems for communication. All animals need a reliable and efficient means of expressing themselves, whether they are communicating their location, engaging in courtship rituals or raising the alarm. We are only just beginning to unravel the complexities of the different methods animals use to exchange information with one another, from the meaningful primate vocalizations described in this chapter to the intricacies of the incredible honeybee communication system and the newly discovered ability of elephants to emit sounds that are detected through their feet.

DANCE LANGUAGE OF HONEYBEES

NATURAL HABITAT
Worldwide

BEES ARE EXCELLENT COMMUNICATORS THROUGH DANCE MOVEMENTS, AND EQUALLY ADEPT AT DECODING THE DANCES. DANCES PERFORMED BY BEES WHEN THEY RETURN TO THE HIVE COMMUNICATE INFORMATION ABOUT THE QUALITY OF A NEW FOOD SOURCE AND TELL THE OTHER BEES EXACTLY WHERE TO FIND IT.

It is hard to imagine that a walk through a garden or city park can allow us to see the start of one of the most complex communication systems in the world. During the warmer months of the year, a closer glance at a flower can provide a glimpse. Watch for bees hovering close to the flower's centre, gathering pollen or nectar to take back to the hive. It is what happens when this loyal worker bee returns that the amazing story of the "dance of the bees" begins.

Look what I found

Bees are social insects that live within a highly organized, complex social structure that has evolved over millions of years. The hive and its inhabitants function very effectively together, resulting in an efficient and orderly system that maximizes the health and well-being of the developing larvae and the all-important queen bee.

One of the most important resources required by the hive community is food, and bees have developed a remarkable communication system that provides information about food quality and how far away the source is, and gives explicit directions for finding the source. On returning to

THE BEES WE NEED

Around the world, bees are critical to the world economy, because a great deal of agricultural success is dependent upon the pollination of many types of food crops. Bees are significant contributors to pollination, and without their assistance, the results can be disastrous.

During some growing seasons over the past decade, diseases have decimated bee populations, and as a consequence, entire crops have failed. More recently, bee colonies are dying out completely, and scientists have not yet determined the cause. It is critical that the cause of the colony collapses be found, so that bees can thrive in their communities and continue to aid our own foraging efforts.

BEES BAND TOGETHER
Busy bees share their workload. Honeybees forage communally by signalling the location of rich sources of nectar to one another.

See also
You can count on an ant to find its way home,
page 130

the hive after finding a new food site, the bee performs an elaborate system of dance movements inside the hive. Keep in the mind that the comb structure inside a beehive is vertical, with the combs hanging parallel to each other from the top of the hive, so the returning bee has to "dance" on a vertical surface.

Dr Karl cracks the code

In 1973 Dr Karl von Frisch shared the 1973 Nobel Prize for Medicine with two other animal behaviourists, Konrad Lorenz and Niko Tinbergen. All are now considered the fathers of the field of ethology, which is the study of animal behaviour, typically in its natural habitat.

During the 1940s, pioneering studies carried out by Von Frisch at the University of Munich identified the amazing abilities of honeybees to provide their hive-mates with important information. After some 6,000 observations of the behaviour patterns produced by the bees, von Frisch had untangled the mystery of the dances, and discovered that information is conveyed through two types of dances: the round dance and the waggle dance.

The round dance

If a food source is less than 100 m (300 ft) from the hive, the returning bee performs a simple round dance, outlining a circle in one direction on the vertical surface of the hive, then reversing the circle in the other direction. On translation of the round dance, bees know that food is close to the hive, and outgoing foraging bees can use other cues, such as colour or odour, to locate the site.

"LISTEN" AND LEARN

Bees must translate the "dance language" provided by the returning forager bee to ascertain in which direction to fly when they leave the hive. If a food source is close to the hive, the returning bee performs a simple round dance. If the food is more than 100 m (300 ft) from the hive then a waggle dance is performed. The waggle portion of the dance provides the other bees with the angle of the sun, relative to the food source and the hive and also the distance of the food source from the hive. This flexible system allows for communication about food locations regardless of the sun's movement throughout the day.

The round dance
The round dance is simpler than the waggle dance, and is performed using circular movements only.

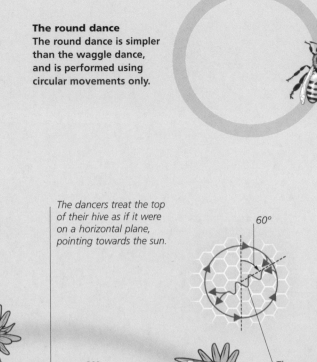

The waggle dance
This diagram shows the highly stereotyped movement patterns of the waggle dance.

If the food source is located in the same direction as the sun, the bee's dance points toward the top of the hive.

Bees adjust the angle of their dance according to the relative positions of the sun, their hive and the food source.

The dancers treat the top of their hive as if it were on a horizontal plane, pointing towards the sun.

The speed of the waggle part of the dance gives information on the distance of the food source from the hive.

If a food source is located in the opposite direction to the sun from the hive, the bees dance vertically downwards.

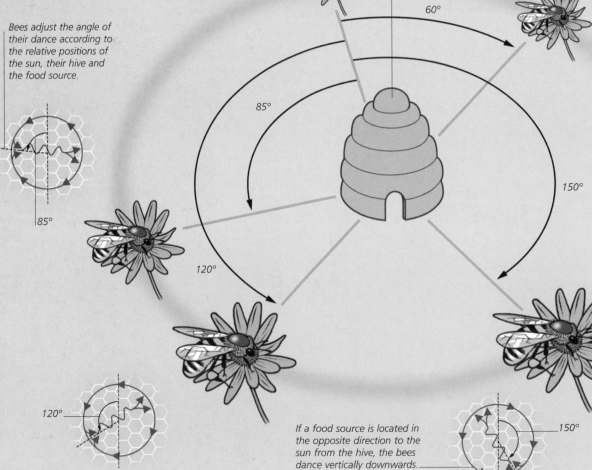

The waggle dance

The waggle dance is a whole different story, that was deciphered only after hundreds of hours of study by von Frisch. The waggle dance itself also has some elements seen in the round dance. It is performed inside the hive, on the vertical surface of the honeycomb, and includes circular movements similar to those of the round dance. However, when the bee completes half a circle, she turns and moves up the comb, bisecting the circle. During the straight run through the circle, the bee wiggles her body back and forth in a dancelike pattern, as if "cha-cha-cha-ing" up the comb. Once at the top of the circle, she turns in the opposite direction, and completes another half a circle, then performs another waggle dance up the comb. Overall, this results in a tight pattern that resembles a squat figure-of-eight.

When the bee performs a waggle dance the onlookers know that the food is some distance from the hive. However, they need to know in which direction to search; information that is also provided by the waggle dance. The bee adjusts the orientation of the waggle dance, based on the position of the sun, and the angle represented between the hive, the sun and the food source. The orientation of the waggle-dance line provides the direction of the food source. However, remember that the comb is vertical. In order for the waggle portion of the dance to convey the position of the sun, regardless of the time of day, the bee performs the waggle dance as though the sun were positioned at the top of the hive. In other words, bees have a built-in GPS system that can be used all day, regardless of the sun's movements. The angle shown as the bee waggles across the circle means that the other bees should fly out of the hive and in the direction represented by the waggle dance. The bees must translate the relative position of the food source, based on the actual position of the sun, even though the waggle portion was performed on a vertical surface.

Perhaps even more significant is the fact that the waggle dance also communicates the distance to the food source, based upon the speed of the waggle portion. The slower the dance, the more distant the food site. For example, if a food source is 10 km (6 miles) from the hive, the bee dances the waggle portion very slowly, approximately eight waggles per minute, but if the food source is only 500 m (1,635 ft) away, the waggle speed increases to about 25 per minute. The onlooking bees are able to compute the distance to the food source, based upon the dancer's speed.

Skill for life

If each generation of the thousands of bees in a hive had to learn each of the dances individually, the bee community could not function as it does today. Instead, the emergence of the bee's dance language is part of its genetic make-up, an adaptation of millions of years of evolutionary change over time. Luckily, their communication system is now "built in", and each forager bee can both perform the dances accurately and understand the information that is being conveyed by other dancing bees.

COMMUNICATION
Information dances are integral to the complex social network of a bee hive.

GROUND SQUIRRELS LOOK OUT FOR THEIR OWN

BELDING'S GROUND SQUIRRELS HAVE A COMPLEX SYSTEM OF VOCAL COMMUNICATION USED TO SPECIFICALLY PROTECT FEMALE RELATIVES FROM AERIAL AND GROUND PREDATORS.

NATURAL HABITAT
Far western USA

Belding's ground squirrels are a highly social species, living in large groups in sub-alpine environments in the far western United States, including meadows, sagebrush areas and forests. Male squirrels, when they reach the right age, will leave their natal group and search out new communities, while the females remain with their natal group for life. This means that all females in the community are related, and these relationships form the basis for altruistic alarm calling that protects the entire social group, and contributes towards the squirrels' genetic continuity.

Trill and whistle

While ground squirrels face many dangers, their most serious predators are weasels, snakes, infanticide by other squirrels and aerial predators such as hawks. Females guard their burrows by using scent marking to identify their territory, and also by chasing away any intruders. Very young ground squirrels are especially vulnerable to predation when they first emerge from their natal burrows, and only 40 to 60 per cent of the pups survive until the autumn. During the rest of the summer, the remaining pups have to dig a burrow for hibernating, and stay alert for predators.

Ground squirrels forage socially, which helps protect them from predation: a larger number of individuals means there is greater vigilance among the group, should a predator be seen. If a predator is spotted, the observing squirrel will stand up and give a particular vocal alarm call. Young ground squirrels are not born with the ability to discriminate between "trill" alarm calls that are usually given when a ground predator is spotted, and an alarm "whistle" that is produced when an avian predator is in sight. It may take up to a week for the young pups to learn the difference between the two types of calls. When the other members of the group hear an alarm call, they either retreat immediately to the safety of their burrows for a whistle or, if the alarm call is a trill, they "post", a bipedal stance, which allows the squirrels to look around and see what prompted the call.

However, not all of the squirrels are equally likely to call; instead it appears that while all females and males will give the whistle alarm, trill alarm calls are only given by females in the

MAKE THE MOST OF THE WEATHER, WHILE IT LASTS

Due to extreme climate changes in their habitat, Belding's ground squirrels are active during the day only for about three to four months of the year, in the summer, when they forage for seeds, flowers and vegetation. Ground squirrels hibernate for seven to eight months of the year, and therefore must eat enough while active to save sufficient fat storage for the long hibernation period. When adults emerge from hibernation, they mate immediately. At that time, some males will disperse to new groups, while others will spend several months gaining weight in preparation for their return to hibernation. A month after mating, the females give birth. Litters of five to eight pups are born, and raised by their mothers in their underground burrows until they are 3 to 6 weeks old. Shortly after, juvenile males leave to join other communities, while females remain with their natal group for life. Approximately a month after the juveniles emerge from their burrows the adult males will go into hibernation.

SPECIFIC CALL

If an aerial or ground predator is spotted, the observer squirrel stands upright and begins to vocalize using specific alarm calls that indicate the type of predator in sight.

presence of squirrels that are related genetically. The whistle call causes all squirrels in the vicinity rush to their burrows and confuses the predator, making it unlikely the caller is caught; making it a "selfish" call. Conversely, the trill alarm is much more risky to the caller; an altruistic call. Because females do not leave the social group they were born into, they are all inter-related. The concentration of female relatives has enhanced the development of altruistic alarm calling, as well as encouraged collaborative defence of territories from outsiders.

Keep it in the close family

There can be a high cost to giving alarm calls, because they identify the location of the caller, therefore making her much more vulnerable to possible attack and subsequent death. However, the long-term benefits to the group are apparent, so females will take the risk to call if very close kin are nearby. Similarly, females are much more likely to take on an intruding squirrel if the invasion occurs within the territory of close kin living nearby, such as sisters or daughters. The likelihood of their defence is diminished if the intrusion is into the home area of more distant relatives, such as aunts or grandmothers, since assistance to closer relatives means that more closely related kin may survive, and therefore pass on their shared genes to subsequent generations.

Who are you?

An interesting question is: how do ground squirrels recognize one another, since this would be necessary in order for differential altruistic calling to occur. There certainly appear to be a great deal of physical similarities among the community members, so other mechanisms must be in place for individual recognition to occur.

The answer would appear to be in odour cues. Ground squirrels produce two different types of odours that aid in recognizing relatives, including those produced from glands in their mouths and a second set of glands on their backs, both of which vary according to how related the individual animals are. When squirrels meet one another they behave as though they are kissing. What they are actually doing is smelling the odour emitted from each others' mouth glands, which allows them to identify one another as relatives – including the degree to which they are related, such as sister, cousin or unrelated – and respond accordingly.

See also

Wild vervet monkey "smart" alarms, *page 68*

I'll scratch your back now, if you'll scratch my back later, *page 170*

WILD VERVET MONKEY "SMART" ALARMS

CLEVER VERVET MONKEYS HAVE SPECIFIC ALARM CALLS FOR THREE DIFFERENT POTENTIAL PREDATORS: LEOPARDS, SNAKES AND MARTIAL EAGLES.

NATURAL HABITAT
Sub-Saharan Africa

The vervet monkey is a medium-sized primate that lives in sub-Saharan Africa. They are considered semi-arboreal, spending part of their time in the trees, and semi-terrestrial, with time spent on the ground playing or feeding throughout the day. They are active principally in the day, and spend the night sleeping in trees, for protection.

The alarm vocalizations of wild vervet monkeys are among the best-studied in the world. Recent playback studies using pre-recorded alarm calls have revealed that the calls they produce when different predators are seen have specific meanings for the other monkeys in their group.

Specific alarm calls

The three predators that vervets must be careful to watch out for are leopards, martial eagles and other raptorial birds, and snakes. Two of their predators live on the ground, and the other is an aerial enemy, so the monkeys must react differently to each of them. If a leopard is seen by a vervet, he or she gives off a specific alarm call, resulting in the other vervets running up into the trees onto smaller branches that cannot support the weight of the big cat. However, if a snake is spotted, the alerted animals will stand up on two legs to search for the snake. Birds that may attack from the air require that the vervets have a different escape strategy. In this case, the monkeys jump into the bushes,

AFRICAN MONKEYS
Vervets are also known as green monkeys, and are small grey primates with a dark border of hair highlighting their faces. They are seen in many countries throughout Africa, including tourist camps where they often get into trouble! Baby vervets are particularly vulnerable to attack by predators.

IMMEDIATE IDENTIFICATION

Many species give alarm calls in response to predators, but vervet monkeys produce different calls for specific predators.

See also
Diana monkeys spread the news, *page 74*

VERVET CALLING

Vervet monkeys not only produce distinctive calls, such as screaming, barking and chattering, but they also respond to these calls in distinctive ways.

EAGLE CALL

If an eagle or other martial bird is seen, the representative alarm call is given by one or more monkeys, and the troop retreats to nearby bushes until the danger has passed.

and remain under cover until the eagle has flown on. All three separate behavioural responses are coordinated with the particular predators in their territory, and each alarm call is correlated with only one of the three predators. This suggests that the alarm calls of vervet monkeys function as primitive, name-like labels for leopards, eagles or snakes, and that a vervet who hears the separate calls will react accordingly.

Learning what to say when

It is assumed that young monkeys are born with certain features of the alarm vocalizations available, but they must still learn to use the calls at the appropriate times, and in response to the correct predators. Sometimes baby monkeys give eagle alarm calls to almost anything from above, such as a falling leaf. Through social learning from other members of the troop, however, the young monkeys learn to produce the right alarm call only when the corresponding predator has been spotted. Therefore, with time and experience, the younger vervets learn to discriminate the relationship between the alarm vocalization and the specific predator. That is, they learn to "say the right thing at the right time". Otherwise, failure to acquire a full understanding of the calls and their referents could have serious repercussions in a life-and-death situation.

A social call

It has been discovered that vervets also use other vocalizations in different types of social interactions, and some may also be referential. That is, like the specificity of the alarm calls for different predators, other vocalizations by vervets appear to convey information that may represent an individual monkey or express the nature of a social interaction.

In another study, researchers recorded alarm calls of young vervets, then hid a speaker near where the vervets spent time on the ground. When an adult female monkey was near the speaker they played the recorded calls. The adult females looked at the mother of the infant whose alarm calls were being played, and not at the infant itself. This suggested to the scientists that adult

MONKEY BACKGROUND

Vervet monkeys live in large groups of some males, numerous females and their offspring, sometimes in numbers up to 80 monkeys, although a more typical group size is about 20 individuals. Vervets inhabit savannah lands and mountain areas and have a highly varied diet that mainly consists of fruit, although they also eat seeds, leaves, insects and even small mammals.

female vervets could recognize the individual calls of each infant, and also knew something about the relationship between the young monkey and its mother.

Studies of other vervet vocalizations yielded similar results in another context. This time, the researchers recorded grunts that the monkeys made in different situations. It had been assumed previously that slight differences in the acoustic features of the grunts had to do with the different context in which they were heard. However, when adult vervets were tested using playback prerecorded grunts, they discovered that the animals' behavioural responses were different and consistent. That is, they responded to the subtle differences in the grunts regardless of the situation or context, suggesting that the grunts may have different meanings themselves. For example, if they played grunts from a male who ranked low on the dominance hierarchy, directed to the highest-ranking male, the monkeys would just sit there. However, if they heard grunts that had been recorded from a dominant male, they moved away quickly from the area of the speaker. Precisely what the different grunts actually mean to the monkeys is unknown, but clearly differences, no matter how slight they may sound to the human ear, are immediately detected by other monkeys.

LEOPARD CALL

The sight of a leopard will provoke a particular alarm call from the vervets, resulting in a mass exodus by the group, high up into the trees and along flimsy branches that can't hold the leopard's weight.

SNAKE CALL

When vervets hear the alarm call for snakes, they stand upright and look around, trying to locate the snake that must be nearby.

BABOONS: KINGS OF EXPRESSION

NATURAL HABITAT
East Africa

BABOONS IN THE WILD HAVE ACCESS TO A COMPLEX COMMUNICATION SYSTEM THAT INCLUDES POSTURES, FACIAL EXPRESSIONS AND VOCALIZATIONS THAT CAN PROVIDE THE LISTENER WITH INFORMATION ABOUT EMOTIONAL INTENSITY, SPECIFICITY OF PREDATORS AND CALLER IDENTITY.

Baboons are able to convey a range of emotional intensity based upon the repetition of some sounds associated with other features, such as facial expressions or postures – their principal form of communication.

Vocal and non-vocal displays

Communication among baboons is considerably complex, and most of what has been learned about the levels of complexity of their vocalizations is quite recent. Some scientists put the number of vocal calls at 30, including a range of screams, barks, grunts and alarm calls. Non-verbal components are also part of their communicative repertoire, including body postures, lip smacking and facial expressions such as yawns revealing enormous canines in males, as an impressive warning display.

Communication among baboons does not always require a vocal component. For instance, baboons can display aggression very effectively with an "open-mouth threat" that consists of raising the eyebrows and revealing the whites of their eyes, followed by baring of the teeth. At greater levels of hostility, the baboon can make its hair stand up, creating the impression of a much larger body size, and also include a threatening vocalization, followed by a strong slap to the ground. Similarly, a reversal of some of the same features can denote submission, in that baboons can produce a "fear face" in response to such an aggressive display by pulling the corners of the mouth back, resembling what in humans is a wide smile. For baboons, however, it means that the recipient of the threat is submitting to the more dominant animal, and indicates they are fearful of an attack.

BELLOWING BABOON
There are five sub-species of baboon. Here, a male hamadryas baboon produces an open-mouthed display that communicates a medium level of threat to his opponents.

Depth of meaning

Recent field studies by primatologists have revealed that the vocalizations of baboons have far more substance than previously thought. A range of studies on several species of baboon have shown that they appear to have flexible and, in some cases, specific types of vocalizations, including some that refer to different types of predators, identify the caller or indicate an animal's emotional state. For example, in female chacma baboons, one of the four savannah-dwelling species, one type of call known as "loud barks" can vary according to the individual caller, the type of predator and the social context at the time of the call. By studying the acoustic features of the calls in great detail, field researchers were able to determine that they had a range of variability along a continuum, with more tonal barks given when a baboon wished to maintain contact with its group, or as a contact call when an infant became separated from its mother. More noisy versions of the call were given as "alarm barks", when large predators were seen. Within the alarm call category there were measurable and significant differences between calls that were made if alligators were seen and those that were produced in response to mammalian carnivores. Both types of alarm call were distinctly different from the types of contact call heard, and there were also quantitative, consistent differences in all types of call for different individuals. These findings show that female chacma baboons learn the caller's identify, and whether the call denotes the presence of a predator or refers to concerns about contact with the group or a specific female and her infant.

Further study

This level of detail regarding baboon vocalizations had not been documented until scientists began to apply similar methods and procedures that had previously been used to study vervet monkey vocalizations. Consequently, many new studies addressed similar questions regarding the complexity of vocal communication in a wide range of species, including other non-human primates. Therefore, we now know that the calls of baboons can provide information about objects and ongoing events, as well as a level of affective intensity in the context. This translates into information about emotions and meaning, either about a specific predator or individual baboon, or some combination of the two. Baboons appear to be exquisitely equipped to listen to exchanges between other individual baboons, and can gain important social information, indicating the level of sophistication inherent in their vocal communication system.

MALE BABOONS
There are often aggressive encounters among males, as they jockey for rank and access to receptive breeding females.

SOCIAL STRUCTURE

Baboons are the largest of the Old World monkeys in Africa, with males weighing anywhere from 15–37 kg (33–82 lb). Baboons show sexual dimorphism whereby males are much larger than females, and the biggest and fiercest adult males dominate the groups. Baboon social structure consists of multiple adult males and about twice as many adult females and their young offspring. Their troops can get exceedingly large, even including hundreds of baboons that manage to maintain fairly consistent order through the complex hierarchies among both males and females. Females have their own ranking system within the core of the group that consists of adult females, juvenile animals and infants, and offspring that share the same social rank as their mothers. Female offspring stay in the troop where they were born (their natal group) for their lifetime. Young males leave the troop as they reach sexual maturity to join other troops. Adult males serve a protective and mediating role over the entire troop.

BABOON BACKGROUND

Of the five different species of baboon, four prefer the savannah and other types of semi-arid habitats, and one species inhabits the cliffs and hills along the Red Sea, dispersing for day foraging, and returning to their cliff homes at night. With respect to diet, baboons are considered "opportunistic omnivores", eating a wide range of foods including grass, which makes up a significant portion of their daily intake; leaves, bark, berries, seeds, flowers, insects and some meat from fish, shellfish, birds, hares and even other monkeys such as vervets or small antelopes. Baboons also forage on tubers that they dig from the hard soil, a resource that is not exploited by other primates, but has provided them with sustenance during seasonal changes of fruit availability. They spend the majority of their time on the ground, rather than in the trees like many of their monkey relatives. However, they do climb trees to eat at times or at night to sleep away from terrestrial predators.

BABOON EXPRESSIONS

Baboons communicate not only through vocalizations, but also by producing distinctive facial expressions indicating a range of emotions of differing intensity.

THREAT STARE

Direct eye contact with tense, closed lips communicates a mild threat.

OPEN-MOUTH THREAT

An open mouth with partially exposed teeth communicates a more intense level of threat.

A FULL "THREAT YAWN"

A full "threat yawn", with the canines clearly displayed, communicates a high level of hostility.

PLAY FACE

Two juvenile olive baboons produce distinctive "play faces" (mouth open with the upper teeth covered by the upper lip) as a signal to chase or start rough-and-tumble play.

See also
Wild vervet monkey "smart" alarms, page 68

DIANA MONKEYS SPREAD THE NEWS

ALERTING OTHERS
The African hornbill takes advantage of warnings by Diana monkeys that a crowned eagle, its main predator, has been seen approaching.

NATURAL HABITAT
Western Africa

PREDATION BY DIFFERENT TYPES OF CREATURES LIVING IN THE SAME HABITAT IS A CONSTANT CONCERN FOR DIANA MONKEYS. IN RESPONSE TO SUCH THREATS, THIS SPECIES OF OLD WORLD MONKEY HAS DEVELOPED MEANINGFUL VOCAL SIGNALS FOR INDIVIDUAL TYPES OF PREDATORS.

The communication system of Diana monkeys features a range of general vocalizations, including alarm and contact calls, body postures and visual cues through facial expressions that are enhanced by their colourful hair patterns – although their bodies are primarily covered with black hair, the monkeys have a distinctive white throat, beard, ruffed collar and a stark white stripe of hair down each arm and thigh. They also have well-developed musculature in the cheeks and around the mouth that contributes to their flexible facial displays.

Exchange of information

Similar to vervet monkeys, Diana monkeys give alarm calls that differentiate between the types of predators seen lurking in the area. There are four principal aggressors for which they must be vigilant: leopards, eagles, chimpanzees and human hunters. Male and female Diana monkeys have acoustically distinct calls for eagles and leopards, but nevertheless coordinate exchange of information. For example, if a female Diana monkey hears a male's alarm call for a leopard, she produces a different but consistent call that she also gives if she hears a leopard's growl. Similarly, if a female hears a male Diana monkey give a completely different alarm call that is only made for an eagle, her vocal reply to the male's eagle call is acoustically the same as the alarm she gives if she hears the shriek of a crowned eagle. If they see chimpanzees that can chase the monkeys through the treetops, or humans who have guns and other weapons that have a long range, Diana monkeys are totally silent, and quietly retreat from the area without giving off any alarms. Presumably, their sudden retreating movements are recognized by the other members of their group, and they, too, move off silently.

MONKEY BACKGROUND

Diana monkeys are medium-sized primates, noted for their striking colouration, that inhabit the upper levels of closed-canopy forests from Sierra Leone to Ghana in West Africa. Their tails can be as long as 74 cm (29 in), with a body range of in 40–55 cm (16–22 in). Diana monkeys are mainly arboreal and active during the day. They have an omnivorous diet, which means that they eat almost anything that is non-toxic, including fruit, leaves, flowers, insects and small invertebrate animals.

In terms of social structure, Diana monkeys live in polygynous, single-male, multi-female social groups of 15 to 30 animals, with the male mating with more than one female in the same breeding period. Their infants have a long period of dependence on maternal care, as is typical for most primate species. Female offspring stay with their mothers their entire lives, while adolescent males leave at sexual maturity to join other communities, thus ensuring a natural form of protection from incest.

See also
Wild vervet monkey "smart" alarms, *page 68*

HAIR HELPS
Diana monkeys have highly distinctive hair patterns that function within their communication system of facial expressions and body postures, in addition to their vocalizations.

Sharing the knowledge

Remarkably, the flexible alarm system of Diana monkeys, including their vocal and behavioural responses, has been successfully coopted by at least two other species that inhabit the same eco-niche. Like Diana monkeys, yellow-casqued hornbills are also vulnerable to attacks by crowned eagles, but not ground predators such as leopards. When tested with pre-recorded monkey alarm calls for eagles and leopards, researchers found that the hornbills responded to them accordingly. That is, the birds reacted by seeking shelter if they heard the recorded call for a crowned eagle, but not a leopard alarm call. They also responded similarly if they heard recordings of actual leopards or eagles, thereby confirming that they could differentiate between both the real calls and the representational vocalizations that Diana monkeys give if either predator is spotted. Consequently, the hornbills' association with Diana monkeys within the same strata of the forest, and capacity to learn the differential alarm calls, allows them to take advantage of the monkeys' advertisement of a potential predator from the sky.

Similarly, red colobus monkeys also share overlapping habitats with Diana monkeys, and it has been determined that the colobus associate with them for increased safety. When Diana monkeys are present, red colobus have been observed to spend more time towards a lower level in the canopy, and although they are more exposed to the forest floor, they do not look down as often to watch for leopards. When the groups forage together and are approached by an observer or an eagle, the Diana monkeys give the vocal alarm call that alerts all the animals. In addition, the type of vigilance observed by the two species differs, but overall provides more protection for both during foraging. Scientists have concluded that red colobus monkeys, like the hornbills, benefit from associating with Diana monkeys during feeding because of the reduced pressure from aerial predation.

WILD CHIMPANZEES: MASTERS OF COMMUNICATION

NEW DISCOVERIES HAVE DOCUMENTED THAT CHIMPANZEES HAVE UNIQUE LONG-CALL VOCALIZATIONS THAT IDENTIFY GROUP MEMBERS OR OTHER CHIMPANZEES, AS WELL AS OTHER DISTINCT VOCALIZATIONS THAT SERVE SPECIFIC COMMUNICATIVE FUNCTIONS.

NATURAL HABITAT
Western and
Central Africa

Community life

Chimpanzees are extremely social and live in communities within a specified home territory that is defended by the adult males of the group. Their social structure is difficult to characterize, but the basic unit that exists within a chimpanzee community is a mother, her infant and other young offspring. Chimpanzees live in a fission–fusion society, which means that animals are coming and going all the time. However, they identify with a specific community, and always return to the group after their forays away. Whenever a chimpanzee or group of chimps returns to the community the chimps become very excited, with hugs and actual kisses going around the group as shrieks of joy and excitement fill the air.

Chimpanzees are highly intelligent, and have probably evolved extraordinary capacities to live within dynamic and challenging communities that require recognizing every member; remembering chimps that have left, sometimes for a considerable period of time; understanding the complexities of the social hierarchy and rule structure of their community; and having the necessary attentional and observational skills to learn the many types of tool use that have now been documented for wild chimpanzees.

Leadership of the group is often comprised of coalitions of related adult males, such as brothers, and perhaps one other adult male with whom the other members of the coalition have a social bond. The dominant males keep the peace, and are also constantly vigilant for predators such as leopards or, their biggest enemy, human intruders. Males make a complete border patrol around the community's territory, on average every four days, looking for possible incursions of males from neighbouring communities who may be seeking females. Chimpanzee mothers

KING OF THE CASTLE
A chimpanzee reaches the climax of the pant-hoot, giving a loud, whooping scream that can be heard far across the treetops.

See also
Wild vervet monkey "smart" alarms,
    page 68
Baboons: kings of expression,
    page 70

do not travel as much as males, and tend to forage within the central portions of their territory. Different from many other primate species, young, sexually mature female chimpanzees leave their natal community, and emigrate to another community, thereby ensuring that incest will be avoided.

Vocalization and identification

Chimpanzees have a remarkable set of communicative strategies at their disposal, including naturally occurring gestures that are, in some cases, highly similar to human hand gestures, as well as facial expressions, body postures and a repertoire of vocalizations that has only just begun to be studied. Field researchers are beginning to find that chimpanzees are not simply producing vocalizations related to emotional states, but rather are able to provide a greater amount of information through their calls than previously thought. In fact, the importance of a chimpanzee's ability to recognize the meaning and identity of vocalizations from group members or strangers may even be a matter of life and death. If a lone male from one community encounters a border patrol, which usually includes five to seven adult males, he may be unable to escape. Field scientists have come upon single dead chimpanzees, and their bodies often show evidence of cooperative killing by other chimpanzees. Consequently, as part of their socialization, young chimpanzees must learn who is who within their group, the nature of what particular vocalizations represent and also who is making the call. All of this requires a big brain, and chimpanzees have the capacity to learn much more than is required to survive in the wild, similar to humans. It is this behavioural flexibility and learning capacity that sets the great apes and humans apart from other non-human primate species.

Long-distance call

As scientists have delved deeper into the potential for chimpanzees' vocalization to include more than just their emotional state, new findings are appearing from field workers conducting studies in several East African sites, including the Kibale National Forest in Uganda, and the Mahale Mountains National Park, located in Tanzania. The chimpanzee's vocal repertoire includes screams, hoots, both loud and soft alarm calls, shrieks, and has been estimated to include at least 15 different vocalizations. One call that has been studied in particular is the "pant-hoot" that can be used for long-distance communication across the territory, but also more flexibly in other social contexts. Pant-hoots are quite impressive, beginning with low-pitched, breathy hoots in a sequence that builds quickly with the addition of hoots of increasingly higher frequency pants that are produced by breaths emitted in and out, with the help of the chimp's diaphragm, vocal cords and a large throat sac that contributes to the amplitude of the call. Finally, the pant-hoot reaches a tremendous crescendo with a loud, high-pitched scream that is often associated with drumming or slapping against trees or other hard surfaces in the immediate area. There is no other chimpanzee vocalization that has this distinctive pattern, including acoustic features that allow other chimps to recognize, among other things, the individual caller.

Both males and females use the vocalization as an expression of general arousal and excitement during interactions with the group, but the call can provide significant information to distant recipients as well. Because they can carry through the forest canopy, pant-hoots are used to stay in contact with other specific chimpanzees, indicating the location of allies or family members, as well as revealing strangers that may be nearby.

Distant communities

Remarkably, in addition to providing individual identities, these vocalizations appear to have both genetic and social learning components. This was revealed when researchers studied the vocalizations of two large populations of males from chimpanzee communities that were quite distant from one another. In a very real sense, these two different chimpanzee groups speak different "dialects", with some overlapping components. Because vocalizations differ depending upon age and sex, only male calls were recorded. The question that naturally arose was whether the observed differences had to do with learning vocal features within a community, the contribution of different habitats with different capacities for carrying sound, or whether genetics played a significant role. Clearly, there are geographic differences, but how might these affect the meaning or function of the calls? These questions could only be answered by rigorous study of the finer acoustic features of each group of chimpanzees.

Because of the long distance between the two groups' habitats, each was probably isolated from one another; therefore something else was contributing to the differences. Several possible conclusions could be drawn. For example, it is possible

DIFFERENT CALLS

Once a large sample of pant-hoots was recorded and context information was evaluated, these analyses revealed that both groups used them for maintaining contact with allies over long distances. In addition, both groups included the rising introduction when they produced pant-hoots. However, there were measurable and significant differences between the two male populations in the speed and duration of the individual elements in the introductory portion of the call. Although it is likely that the identity of callers could be heard in each group, the Ugandan chimps had longer elements at slower rates as they began their pant-hoots. But the Tanzanian group had a longer overall introductory portion, making it possible for even the researchers to easily learn to recognize which calls came from chimps at which site. Overall, calls differed between the groups with respect to individual elements in the calls, the calling rate and the length of each of the pant-hoot phases.

CONCENTRATED CALL
A young chimpanzee is engrossed in the process of producing the hooting part of its pant-hoot call.

that there are genetic differences between the two populations that are affecting the calls, or that differences in their habitats were impacting on transmission of pant-hoots. A forest environment like Kibale can cause the sound to scatter, but if calls are made at lower frequencies and lower rates, they carry better. Consequently, if chimpanzees live in more open areas, like the Mahale community that was studied, calls would not be subject to the same level of interference, and calls with a higher rate and frequency would be sufficient to ensure the pant-hoot was heard at a distance. The study also revealed that the Uganda site at Kibale had a much greater density of other primates living within the habitat, providing yet another factor that can contribute to degrading a long-distance call. Again, the slower rate and lower frequencies of the chimps living there would compensate for further scattering of the sound. Even differences in body size between the two groups may add to subtle differences in sound production, since the Mahale chimps are smaller than those living in Kibale. It seems that all these factors may be impacting the pant-hoots produced by the different communities, including the habitat acoustics, body size and the overall sound environment with respect to population density of all species living within the area.

Future work

Clearly, more studies will be necessary to fully understand the nature and characteristics of chimpanzee vocalizations; the pant-hoot is only one among their repertoire. There may also be contributions from learning, with young males modelling the chimps' developing calls after specific males, likely the dominant animals, in their own communities.

Now that scientists have turned their attention to exploring the vocalizations of our closest living relative, they are uncovering greater complexity and flexibility in their calls, and raising a host of intriguing questions related to chimpanzee vocalizations and the emergence of human speech.

SIGNATURE WHISTLES IN DOLPHINS

NATURAL HABITAT
Worldwide, mostly in the shallower seas of the continental shelves

THERE IS NO DOUBT THAT BOTTLENOSE DOLPHINS PRODUCE DISTINCTIVE WHISTLE SOUNDS UNDERWATER, BUT WHETHER OR NOT INDIVIDUAL DOLPHINS HAVE THEIR OWN WHISTLES THAT ARE RECOGNIZABLE TO OTHER RELATED DOLPHINS IN THEIR POD IS STILL UNDER DEBATE.

Dolphins are found in all the oceans and seas, as well as many major river systems throughout the world. They live in a variety of water types, including salt water, fresh water and brackish water (a mixture of fresh- and salt water). They can be found principally in more shallow waters, but also live in coastal areas and the open ocean.

Bottlenose dolphins show a preference for warmer, shallow water, which is why they are often found along coastlines, including bays that extend inland, and in harbours. The Atlantic bottlenose dolphin is the one species of dolphin that has been studied by scientists, in particular in relation to the sounds they make underwater.

Sound pictures

Murky water often makes visibility extremely difficult for dolphins, so they use a special system for locating objects and potential obstructions underwater. This type of sensory processing is called "echolocation", and is based on signals that emanate from the dolphin's head, specifically an area known as the "melon". The sound signals project forwards, moving through the water, and bounce off protruding rocks, fish or other objects – animate or inanimate – providing the dolphin with feedback about the object's density and size, and how far away it is. Through a complex transformation system that is still not understood, the dolphin receives a kind of "sound picture" of what is ahead, and responds accordingly.

Vocalizing

In addition to their use of echolocation, dolphins also communicate with one another through a system of vocalizations – combinations of sounds acquired in infancy. One of the most studied of these is their "signature whistle". Prompted by early studies of these vocalizations, scientists suggested some time ago

MAKING CONTACT
When separated from their group, dolphins emit contact calls that some researchers believe are unique "signature whistles" that may function to identify individual dolphins.

PRINCIPLES OF ECHOLOCATION

Dolphins use echolocation to find prey and recognize other objects and potential obstructions in their underwater environment, where vision can be limited. Clicks are emitted from an area on the top of the head called the "melon", and bounce off objects in the immediate vicinity of the dolphin. This provides the dolphin with feedback that translates into some form of mental representation of the objects and contours in the area. Precisely how echolocation works as a sensory process is not yet understood.

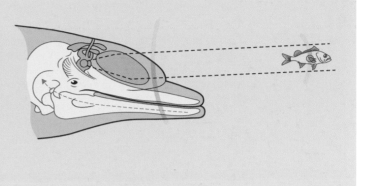

that each dolphin might have its own unique signature whistle, and that it might be used between dolphins for identification of group members or relatives. The hypothesis of individual signature whistles in dolphins has been a contentious one, and has not been completely resolved, despite a significant number of studies done with captive populations, wild dolphins and/or comparisons of vocalizations produced by groups of dolphins raised in each environment.

"Playback" trials

One of the most recent experiments has concluded that dolphin signature whistles do convey specific identifying information, and media reports of the findings have gone so far as to say that dolphins have "names" for themselves, in the form of signature whistles. The research team hoped to tease apart the contribution of voice recognition as the basis for other dolphins recognizing and indicating familiarity to a particular signature whistle. To test this, they removed the specific features of the caller's voice through electronic means, leaving only the basic pitch contours. This meant that the whistle no longer had acoustic features that allowed a listener to identify an individual simply by their "tone of voice". Instead, the listener would have to recognize the overall acoustic configuration in terms of frequency changes before they would be able to recognize the whistle. The altered whistles were played through an underwater speaker, and in the majority of test trials the dolphins would turn towards the speaker if they heard an electronic recording that was similar to a close relative's whistle. This type of procedure is known as a "playback study", because the experimenters use pre-recorded vocalizations, and then present them to individual animals to see how they respond behaviourally. In this case, dolphins oriented towards the speaker, which may mean that they recognized the altered version of a familiar vocalization as that of a relative.

These were important findings, but they did not tell the researchers that the dolphins recognized a one-to-one correspondence with a particular whistle and a specific dolphin. Instead, the dolphins may have demonstrated that they did not need the individualized voice features of a signature whistle to indicate that they recognized it. If young dolphins learn their own signature whistle at a very young age, and they associate it with

THE DOLPHIN POD

The composition of groups among wild dolphins (such as the Spinner dolphins shown here) depends upon age, sex of the animals, their reproductive condition, kinship and previous associations and relationships. Their social organization is fluid, and studies have shown that groups can change members quite often. While they typically swim in pods of two to 15 dolphins, they may join other groups temporarily, forming one very large group, sometimes for only minutes or perhaps hours, and may change group members. Group sizes tend to be correlated with the openness of the habitat as well as the water depth, and these factors are probably correlated with the availability of food resources, foraging techniques and the need for protection from some species of whale and shark that prey on dolphins.

Female dolphins and their calves are closely bonded, and stay together for many years. Young females may return to their natal group, to their mothers or other female relatives, to raise their own calves. Consequently, it is not unusual for dolphin pods to be multi-generational.

Adult male pair-bonds are also very strong and can last for decades. Males form tight coalitions, and cooperate in group defence, as well as a number of different feeding strategies that require them to orchestrate their behaviours for trapping fish. Relationships formed between adult males and females, however, are short, and represent consort breeding relationships only.

their mother and older siblings in a group, it is more likely that they will use components of the whistles they hear produced by these close relatives. Consequently, there may be specific acoustic features that are shared among family members, so if a dolphin heard any whistle that had been recorded from a close relative, especially if the voice itself was altered, it might be responding to a shared feature, more like a family's "last name", if you will, rather than identifying an individual animal.

More work to do

Historically, the validity of signature whistles in dolphins has sometimes been accepted and at other times rejected, based on experimental evidence reported in support of one viewpoint. The next study, from another research team, might publish results that lead them to conclude that dolphins do not have specific signature whistles at all. This process is the nature of scientific investigation, and not until all possible experimental approaches have been explored, along with every potential hypothesis that might account for what signature whistles do or do not represent, will the scientific community accept one position or the other as more likely.

In another study of signature whistles by a different group of scientists, the conclusions were entirely different, after studying what they said were dolphins' "contact" or "cohesion" calls, a category of vocalizations that other scientists have claimed represent individual signature whistles that identify specific dolphins. The team used the same methods and procedures that had been reported by the experimenters who made the

See also
Wild vervet monkey "smart"
   alarms, *page 68*
Baboons: kings of expression,
   page 70

original claims about signature whistles, but found that isolated dolphins produced essentially the same predominant and shared type of whistle, rather than individualized vocalizations that could be ascribed to a particular dolphin. Other studies have found no evidence for signature whistles whatsoever, so the question still remains as to whether or not the vocal repertoire of dolphins includes calls that have specific meanings, or are representational – that they "stand for" one dolphin or another, and therefore have some abstract qualities that would suggest that they share some features of naming, as in human language.

Whether or not the cognitive capacity of the bottlenose dolphin extends to semantic use of vocalizations that refer to individual animals is a fascinating question that will no doubt provoke continued research well into the future. As it stands now, very many species of wild dolphins have not even been studied at all; most research has focused on the Atlantic bottlenose dolphin discussed here. What has been learned about just this one species of dolphin clearly suggests that there is a wealth of information and exciting work to be done, extending these studies to the many other dolphin species around the world.

WHALES CAN REALLY CARRY A TUNE

NATURAL HABITAT
Worldwide

WHALES USE LOW-FREQUENCY, LOW-AMPLITUDE VOCALIZATIONS THAT CAN BE HEARD OVER EXTREMELY LONG DISTANCES, AND THAT HAVE BEEN HYPOTHESIZED TO SERVE AS A COMMUNICATION SYSTEM DURING MATING, MIGRATION AND FEEDING.

Across the numerous species of whales, it has been suggested that the system of song vocalizations they produce is among the most complex of non-human acoustic repertoires heard anywhere in the animal kingdom. The songs consist of a series of repeated sounds that is detectable over long distances, and produced by males only, typically during the breeding season and during the annual migration of some species of whales, such as the humpback whale. Whale songs also occur during the late summer and early fall in feeding areas where the whales gorge themselves on small fish. There also appear to be different types of songs produced by differing whale species, as well as distinct regional "dialects" of songs from whales of the same species that live in different parts of the ocean. Numerous functions have been proposed and tested with groups of whales in various locations, and it is believed that the songs can serve to keep whales apprised of one another's locations during feeding or advertise a male's presence to females in the area.

Song characteristics

Some of the most well-described songs are those of the humpback whale, which were first documented by scientists in 1971, although the sounds had been identified as whale song by the US Navy two decades earlier. Their songs comprise of a variety of tones, from very high-frequency squeaking sounds, all the way to very low-frequency sounds more like a lion's roar. Songs are usually repeated over and over during a single session comprised of a series of shorter songs that are represented by five to seven units or "themes" that each occur numerous times. The entire song can range from five to 30 minutes, with individual themes repeated sequentially over periods that may be hours long.

Blue whales, on the other hand, have entirely different song characteristics. Their songs are used principally by males to attract female mates, and consist of short alternations of two separate tones that last about 20 seconds and are extremely loud. Because

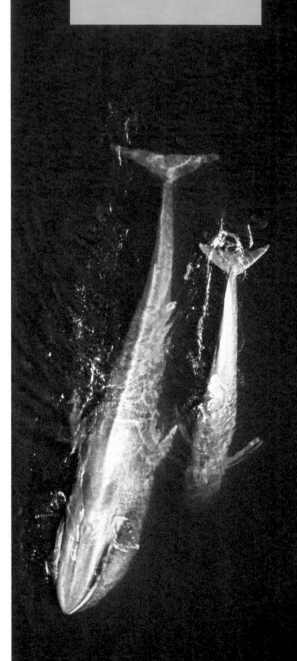

THE RIGHT CHOICE
By choosing a healthy and large male based on his song, female whales optimize their chances of giving birth to strong and healthy calves.

See also
Can elephants hear through their feet?, *page 86*

GROUP SINGALONG
Whales live in different populations throughout the world, and within a group, sing the same song, despite its ever-changing complexity.

conditions for seeing underwater in the murky depths of the ocean are limited, the use of auditory cues by males to advertise their size and overall health allows prospective females to choose optimally for greater reproductive success. In turn, a female blue whale's choice of a male of greater prowess contributes towards her giving birth to a strong and healthy calf.

Blue whales also produce a different song when other whales in their group are nearby. These are known as downsweep calls, and consist of short songs that consist of a single note about five seconds long. Downsweep songs function to maintain contact with other whales that are feeding in the vicinity, and thus are both functionally and acoustically different from mating songs.

Dialects

It is the blue whales that demonstrate distinct dialects among groups that live in different parts of the world. Whales that live in warmer, more shallow, coastal waters, such as off California, have more complex songs. It has been suggested that because they tend to stay closer to one another, the songs do not have to carry as far, and therefore can be more elaborate. In deeper waters, their songs are different, as well as in different areas of the ocean, such as off the coast of South America.

Unanswered questions

Little evidence is currently available as to how whale song is produced and propagated. Interestingly, however, whales have been observed to position themselves with their head at about a 45-degree angle some 20–25 m (66–82 ft) deep in the water while singing. Whales have not been observed in such positions unless they are singing, so there appears to be some relationship between the two behaviours. It has been suggested that somehow this position allows the whale's song to be produced best, or that this stance allows for optimal sound propagation through the water to its intended targets.

Whale song is not static, but instead it gradually changes its features over time. In fact, entirely different sounds and patterns form every year, and are slowly added to the old song. Some patterns from the original song are replaced, and the overall song changes while it is being sung. Mysteriously, these changes occur within the whole whale population, and may change completely within a period of years. Consequently, a song from a particular group of whales may sound entirely different when recorded several years apart. One documented song changed entirely in only two years. Yet all whales in a population persist at singing the same version of their song at the same time. This can include whales living thousands of miles from one another, who continue to sing population-wide songs. How these changes occur and radiate remains unknown. Nonetheless, the eerie beauty and power of whale song remains, as do many questions related to its emergence, explicit functions and distribution among the great whale species of the world's vast oceans.

CAN ELEPHANTS HEAR THROUGH THEIR FEET?

ELEPHANTS CAN CHOOSE TO TRAVEL SOME DISTANCE FROM THEIR GROUP, AND IT WOULD APPEAR THAT FEMALE ELEPHANTS KEEP IN CONTACT USING LOW-FREQUENCY SOUNDS THAT CARRY THROUGH THE GROUND.

NATURAL HABITAT
Africa, India, Sri Lanka, Southeast Asia, Malaysia, Indonesia and southern China

Elephants live a long life in social groups of 20 to 100 individuals. Their groups are composed mainly of related females, including sisters, mothers, aunts and other relatives, their offspring, young males and sometimes totally unrelated elephants. Individual social groups are also organized into larger "bands", with all groups of a particular band living within a larger territory. The overall social structure of elephants is quite dynamic, with animals coming together or moving apart flexibly, although female family members and their offspring that are not yet independent typically remain as a social unit.

While it has long been known that elephants produce a range of vocalizations that include screams and trumpeting, within the last 25 years scientists have discovered that female elephants communicate over long distances through rumbles, or infrasound; very low-frequency vocalizations below the levels that humans can hear. Infrasound rumbles are produced by the elephant's vocal cords in their larynx, and they also have the sensory capacity to hear these low-frequency sounds, although the extent to which infrasound functions communicatively in their community is still being studied in the wild.

Using infrasound

Sound is measured in hertz (Hz), with the infrasonic range between 1 and 20 Hz. Elephant infrasound could best be described as "rumbles" at between 14 and 30 Hz. Rumbles can carry several miles, and can usually be detected over a 50-km (30-mile) range for savannah elephants. Forest elephants have been recorded at even lower frequencies, with 5 Hz the lowest infrasound measured to date. Under the right atmospheric conditions, elephant infrasound has been recorded over 10 km (6 miles) away from its source, and covering an astonishing range of 100 sq km (38 square miles). Elephant rumblings propagate seismic waves through the ground as well as through the air, which helps to further extend the animals' range of

SOCIAL STRUCTURE

An elephant's social group is led by an older, more experienced dominant matriarch. Female elephants stay with their natal group for life, and develop strong social bonds with other group members. Once young males reach adolescence, they leave the group to join herds of other young bachelor male elephants. Adult males seldom join groups. Older males often travel and forage alone, as they spend a significant portion of each year in a condition known as "musth". Adult bull elephants in musth are responding to a surge of testosterone and other reproductive hormones. These hormonal changes result in significant behavioural changes that render bull elephants extremely irritable, aggressive and very dangerous. While in musth, males are seeking receptive females for mating.

Females give birth to their first calf before they are 20, and similarly, young males reach sexual maturity between 12 and 13 years of age. However, such young males rarely have access to breeding females, since they are dominated during sexual receptivity by older and much stronger bull elephants. Typically, it is not until a male elephant's late 20s that he has the necessary strength and size to compete successfully for breeding against other males.

RECEIVING INFRASOUND

Elephants not only use their ears to detect infrasound, but they also use their feet and trunks as receptors for incoming sound.

See also
Whales can really carry a tune, *page 84*

communication. Because infrasound can travel such long distances, it represents a critical way for elephants that normally range over wide areas to maintain contact or communicate the threat of potential danger to other elephant groups or individuals in the area. It appears that some information about the caller's identify is also provided, which would, in turn, allow for an elephant to maintain contact with another specific elephant. However, how such identification mechanisms might be used socially remains unknown.

What is also undetermined is precisely how elephants are able to detect infrasound, in terms of the perceptual processing necessary for decoding sounds at such low frequencies. Similarly, precisely how elephants produce infrasound rumbles also remains under investigation by different teams of scientists around the world. One hypothesis is that some features or components of infrasound are transmitted through the ground, and detected by elephants some distance way through their feet. In support of this idea, elephants have been observed to lean forwards, shifting their weight on their feet, and sometimes to even lay their trunks on the ground, just prior to measured infrasounds being recorded by researchers just adjacent to the animals. Elephants' huge ears can serve as parabolic receptors as well, so that infrasound through the air can be processed auditorily. In a very real sense, elephant communication is "top down" through their ears, and "bottom up", through their feet, simultaneously. The rich findings of the use of low-frequency sounds among elephants has also led to speculation by some scientists that perhaps dinosaurs may have been able to produce and detect infrasound.

INFRASOUND IN A LARGER CONTEXT

Infrasound can occur naturally, for example storms that include thunder or strong winds, and ocean movement can produce these low-frequency sound waves. Jet aeroplanes, factories, engines and many types of explosions also emit infrasound, as do climate or geologic changes around the world, such as volcanoes, earthquakes and tsunamis.

Elephants are not the only animals that can detect infrasound; squid, lions, giraffes, whales, alligators, some species of birds and the rhinoceros, among other species, are also capable of hearing infrasound. Some species hear infrasound through the ground, or seismically, while others detect it through the air, or atmospherically. Still others, such as whales, can process infrasound through water.

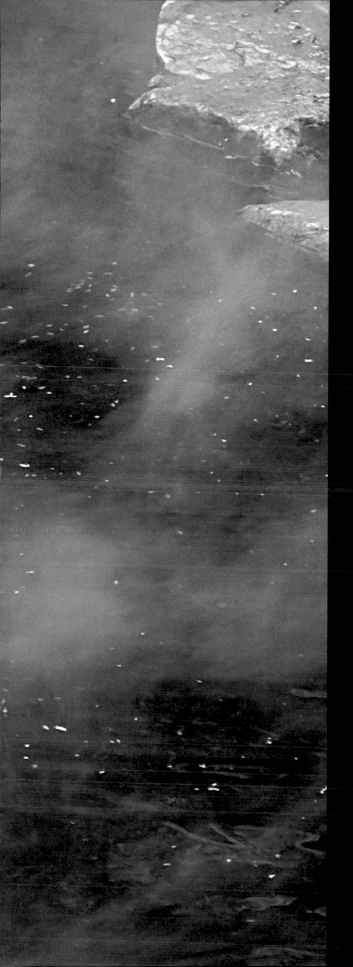

CHAPTER THREE
IMITATION AND SOCIAL LEARNING

The capacity for learning among animals has long fascinated scholars with interests in animal behaviour. Because early studies of animal learning sought to explain learning in humans, processes such as imitation and other forms of social learning that are highly developed in our species have been explored in many other animals under experimental conditions. This chapter examines the emergence of socially acquired behaviours among wild populations of non-human primates, as well as more controlled studies of imitation and similar methods of learning in monkeys, orang-utans, gorillas, chimpanzees, young children, and birds such as starlings, ravens, parrots and Japanese quail.

PLAY TECHNIQUES
Forms of play, such as snowball-making, have spread through groups of Japanese monkeys.

MONKEY SEE, MONKEY DO?

JAPANESE MACAQUES EXHIBIT SEVERAL FORMS OF BEHAVIOUR THAT SEEM TO HAVE BEEN SOCIALLY LEARNED AND THEN PASSED ON FROM ONE GENERATION TO THE NEXT. SOME SCIENTISTS HAVE ARGUED THAT THESE HABITS REPRESENT A SIMPLE FORM OF CULTURE.

NATURAL HABITAT
Japan

Is the old adage "monkey see, monkey do", true? In other words, do monkeys actually watch the clever problem-solving techniques of others in their group and then copy their behaviour? In the 1950s Japanese scientists suggested that not only were the Japanese macaques, which they were studying, capable of social learning, but that they also enacted a simple form of cultural transmission. A new habit was observed as it emerged and spread through the group and, subsequently, it has been maintained down through generations.

Washing and skimming
The researchers studying a group of monkeys, on the remote island of Koshima, tried to entice the monkeys out into the open by scattering pieces of sweet potato onto a sandy beach. The monkeys would sit for a considerable time rubbing the sand off the potatoes before eating

them. One day a low-ranking juvenile female called Imo was seen carrying her potatoes to a freshwater stream where she washed the sand off in the water. One by one, Imo's friends and family began to wash their potatoes, too. Then Imo had a new idea. She began to wash her potatoes in the sea, presumably because it made them taste nice and salty. Eventually, the majority of Imo's group also learned to wash their potatoes in the sea. Even today, when all the original potato-washers are long gone, their descendents wash sweet potatoes in the sea whenever they get a chance to eat them.

Other apparent cultural habits were observed in the Koshima monkeys. After learning to wash their potatoes, the monkeys might literally eat and run. In order to encourage them to linger on the beach, the researchers scattered wheat on the sand. At first the monkeys painstakingly picked up the wheat, one grain at a time. Then

LEARNING BY IMITATING

Despite the old adage "monkey see, monkey do", there is in fact very little convincing evidence that monkeys are capable of imitative learning.

THE "LIGHTBULB MOMENT"

When humans learn from one another, they tend to appreciate the purpose of someone else's behaviour. We often have what one might call a "lightbulb moment" when we see another individual solve a problem and exclaim to ourselves, "What a good idea! I'll do that!" In fact, it has been argued that the emergence of a behavioural habit cannot be considered cultural unless it is based upon imitative learning, which is defined as the faithful copying of the actions of another individual within the same set of circumstances, thereby implying an understanding of its underlying goal or purpose. It is extremely difficult to determine whether the Koshima monkeys learned by imitation or by their own independent discovery while accompanying others in the group to the water at feeding time.

See also

Sea sponges provide padded protection,
page 26

Imo discovered that if she scooped up a large handful of sand and wheat together and dropped it into the sea, the sand would sink while the floating wheat could be easily skimmed off the surface. Just as before, the technique spread through the group.

Japanese macaques live at the furthest northern range of all monkeys, and in winter often have to forage for food in deep snow. One group of monkeys has learned to bathe in natural hot springs on the coldest days. They lounge in the warm water, playing and grooming one another. Just as with potato washing, the habit has been maintained over several generations.

Even some forms of play have spread, and been maintained, in groups of Japanese monkeys. One researcher observed young monkeys playing with stones. Most of the youngsters in the group liked to form large piles of stones just for fun. In another group, the young monkeys enjoyed making snowballs. They formed large balls and carried them around but never actually got as far as throwing them. All these practices – the potato washing, wheat skimming, piling up stones and making snowballs – were learned by one or two individuals and then spread to other members of the group. But can that be called cultural?

Some scientists have argued that the monkeys could have learned these practices independently rather than by copying one another. Simply being exposed to the same environmental factors could have provided each monkey with all the information required to learn such behaviour. Other researchers provided capuchin monkeys and crab-eating macaques with dirty pieces of fruit and a bowl of water. In the course of play they, too, learned to wash the food in the water, suggesting that such behaviour does not depend on social learning.

WARM BATH
When foraging in snow, these macaques have learned to keep warm in the nearby hot springs.

JUNGLE COPYCATS

NATURAL HABITAT

Capuchins: Central and South America

Japanese Macaques: Japan

Marmosets: Brazil

THERE HAVE BEEN MANY ATTEMPTS TO TEST WHETHER MONKEYS CAN IMITATE, FAITHFULLY COPYING THE ACTIONS OF OTHERS. MOST EXPERIMENTS HAVE PRODUCED WEAK OR NO EVIDENCE FOR SUCH ABILITIES.

One way to test for imitation in monkeys is to teach them a copying game, and that's exactly what happened when researchers tried to teach three young capuchins to play a version of "Simon says". The capuchins were being trained to act as helpers for quadriplegic patients and were used to interacting with humans and manipulating objects.

When copying isn't copying

At first the monkeys were taught, by being given edible treats as a reward, to match the actions of a human demonstrator, for example undoing a zip, putting a stick in a cylinder and touching their own arm. But despite the reward system, the monkeys only clearly matched the demonstrated actions from 4–20 per cent of the time. The best-performing monkey then moved to the next phase in which novel actions were demonstrated, but managed a success rate of just 12.5 per cent, turning a screw with a screwdriver, putting two notched blocks together and turning a crank handle. Even then the imitations were crude; rather than copying the actions of the demonstrator, the monkey seemed to be learning how objects could be put together and moved, a type of social learning called "emulation".

Researchers seemed to have more success when they tried a version of "Simon says" with four Japanese macaques. They started by rewarding the monkeys when they simply paid attention, which is obviously crucial because without that there can't be any copying. The monkeys then learned to touch the same place and object as the human, on command. The experimenters claimed that one or two of the monkeys went on to imitate tongue protrusion, touching one's nose, touching one's ear, forming a fist, forming a fist with the thumb extended and pulling a piece of cotton batting in two. Although these results

CONTAGION AND IMITATION

Experiments were conducted that involved presenting a box to common marmosets. One marmoset "demonstrator" was trained to open the box, while a second "demonstrator" would only touch the box if it was covered in honey, whereupon she licked its surface. None of the observing marmosets learned to open the box, and touched it with their hands more than their mouths. In contrast, the group that saw the demonstrator lick the box used their mouths more than their hands.

Other researchers found similar hand or mouth copying in marmosets. They presented marmosets with a cylindrical canister with a pressure lid. A marmoset "demonstrator" either used its hands or mouth to remove the lid and reach the mealworms within. Those observer monkeys who saw the "demonstrator" use its hands, used their hands more than the mouth, while those that saw the "demonstrator" use its mouth, used their mouth more than their hands.

Although these results may suggest that marmosets will imitate some simple aspects of the actions of others, such as hand or mouth use, there is possibly a simpler explanation. Watching another individual use its mouth on an object may automatically trigger feeding behaviour in the observer, so that it uses the mouth more than the hands. The automatic triggering of behaviour via observation of another is called "contagion".

So, despite a great deal of research, most experiments have found very little convincing evidence of a monkey's ability to imitate. Learning by observing others is possible, but monkeys seem to learn more about the way objects are combined or moved (emulation) than by a demonstrator (imitation).

SOCIAL CLIMBER
Experiments have shown that
common marmosets can socially
learn hand versus mouth use
when solving puzzles.

See also
Birdbrained is best, *page 96*

seem impressive, the monkeys' copying movements were often imprecise and they were intermingled with many seemingly unrelated, random actions. Without a rigorous scientific test, it is difficult to tell whether some, if not all, of the imitations were more in the nature of random responding.

Emulation and imitation

"Simon says" games involve imitating actions that have no particular function beyond the copying game itself. Researchers therefore wondered whether monkeys perform better when they are shown actions that help solve a problem, such as opening a box that contains food. They presented common marmosets with a box that had a flap door that could be lifted or pushed to reach a tasty morsel within. When six monkeys were individually given the box, five of them preferred pushing the door and one preferred lifting. The monkey who preferred lifting then became a "demonstrator" to the other five, and three of the group subsequently started to lift the door. However, it is possible that

instead of imitating the lifting action, the monkeys actually learned about the direction in which the door moved, and emulated rather than imitated.

In another experiment, researchers showed a puzzle box to 11 hand-raised capuchin monkeys. The box had a hinged lid that was locked shut by a pair of smooth rods held by a bracket. In order to open the box and reach the treat inside, the rods had to be removed from the bracket. Five monkeys saw the rods being pulled and twisted out, while six saw them being poked out. The monkeys did not learn to copy the demonstrated actions but they did learn to remove the rods in the same direction shown. Again the results were consistent with emulation, not imitation.

NATURAL HABITAT

Parrots: Most warm and tropical regions

Ravens: Europe and North America

Japanese quail: East Asia

Starlings: Europe, north America, Asia, Africa, northern Australia and the islands of the tropical Pacific

Budgies: Australia

BIRDBRAINED IS BEST

WE KNOW THAT MANY SPECIES OF BIRDS ARE EXCELLENT VOCAL MIMICS. SURPRISINGLY, RECENT RESEARCH SUGGESTS THAT THEY MAY ALSO BE ABLE TO IMITATE THE ACTIONS OF OTHERS BOTH IN "SIMON SAYS"-LIKE COPYING GAMES, AND WHEN SOLVING PUZZLES FOR GAINING FOOD.

European starlings have caused plenty of confusion with their ability to mimic the ringtone of a phone. In fact, many species of birds are excellent vocal mimics. Parrots are renowned for it, but researchers have shown that they may also possess equally impressive abilities to imitate the actions of others, and not just apparently functionless actions.

Budgies, ravens and quails

When researchers gave budgies a problem – how to reach food in a dish covered by a paper lid – they trained three parrots to act as "demonstrators". Each parrot removed the lid using a different method. One used its foot, the other used its beak to lift the lid and the third used its beak to nudge it off. Each of the three different groups of observer birds watched one particular demonstration. There was no difference between the methods used by the groups that saw nudging versus lifting with the beak. However, there was a significant difference in beak versus foot use between the groups consistent with the demonstrated methods so, on some level, the birds were imitating rather than emulating. The test was particularly clever because it allowed the researchers to distinguish imitation from emulation. All the birds saw the lid move in the same way; the main difference between the three types of demonstrations was the action used.

Other researchers attempted to use a similar method to test for social learning in semi-wild, hand-raised ravens. A piece of meat was placed in a box with a sliding lid that was opened either by pulling a strap or levering the lid open from the top of

OKICHORO'S ANTICS

A researcher raised an African grey parrot, called Okichoro, in Canada. Oki was kept in a room where he was continuously monitored on video camera and regularly visited by his human keepers. They would enter the room, announce an action, demonstrate it and immediately leave so that they could avoid inadvertently shaping the parrot's behaviour in any way. While all alone in the room, Oki vocally began to imitate the announced actions and later imitated them. The actions included protruding his tongue, nodding his head up and down, shaking his head from side to side and turning 180 degrees. He would also call out "Ciao", followed by slowly waving his foot or wing in an arc.

Some of Oki's imitations echoed unintended actions by his keepers. When the experimenters tried to leave the room, the parrot would often fly up and perch on top of the door. Since this was dangerous, the researchers would bang on the door shouting "Get off!" Eventually, Oki began to shout "Get off", and then started banging his head against the walls or door. And, occasionally, when feeding him peanuts, the keeper would drop one and look down and say, "Oops". Oki began to raise his foot in an apparent grasping motion and then call out "Oops", followed by peering down. Finally, not only would he wave his foot or wing after calling out "Ciao", but he would also then raise and partially rotate his foot in mid-air (as if turning an invisible doorknob), make a few steps, turn to face the other way, and raise and rotate his foot again while vocally imitating the sound of the door clicking as it closed!

WATCH AND LEARN

European starlings are wonderful vocal mimics. However, experiments have shown that they can also learn from watching the actions of others.

the box. Ravens that saw a trained raven demonstrator use the strap also tended to use the strap, while ravens that saw the demonstrator lever the lid preferred the lever. However, the ravens were not necessarily imitating the actions because the demonstrator touched different parts of the box when opening it. Having seen the demonstrator manipulate the strap or the edge of the lid, the birds might well have targeted these sites – a social learning process called "stimulus enhancement" – thereby learning to lever or pull a strap. They might not have been directly copying.

Scientists studying the Japanese quail began by ensuring that the birds' attention was drawn to the same location, with the only difference being the actions used by the "demonstrator". The quail were presented with a lever that had to be lowered in order to reveal a food tray. Those birds that had watched a "demonstrator" step on the lever tended to use their feet, while those birds that had seen a "demonstrator" use its beak, preferred to use their beaks. Although it is clear that the quail socially learned, it is still difficult to tell whether the quails' behaviour constituted imitation. Many birds will automatically begin to peck on seeing another bird peck and will also contagiously flap their wings, thereby increasing the chances of them inadvertently landing on and lowering a lever.

"The ghost condition"

Researchers tried to get starlings to remove a plug from a hole in a platform to reveal a dish of food below. Those birds that saw the plug removed by being pulled upwards tended to pull at it, while those that saw the plug being pushed downwards tended to push. Although their behaviour was consistent with imitation, the birds could just as easily have learned to reproduce the direction in which the plug moved (by emulation) rather than actions of the demonstrator (by imitation). Other scientists came up with an ingenious method for distinguishing between emulation and imitation. They called it "the ghost condition".

Observer birds saw another bird housed in an adjacent compartment to their own that contained the plug platform. Here, the plug appeared to rise upwards or move downwards under its own power, as if being pulled or pushed by a ghost (it was actually moved by an invisible wire). If the birds were able to learn by emulation, then the movement of the object (independent of the actions of a demonstrator) should be sufficient for them to learn one method rather than another. In fact, the starlings only learned pushing or pulling when another bird demonstrated the actions, which lends support to the suggestion that they were learning by imitation rather than by emulation. There is therefore some evidence to suggest that birds are not only capable of reproducing the way they see an object move but of learning to copy the actions performed by others.

SOCIAL LEARNING IN STARLINGS

Researchers tested for imitation in starlings by presenting them with a box containing food. To reach the food the birds had to remove a plug from a hole.

1 | The group of starlings who saw a demonstrator bird pull the plug up and out of the hole learned to pull.

2 | The group of starlings who saw a demonstrator bird push the plug down through the hole learned to push.

3 | Two more groups received "ghost conditions" in which the plug was either pulled up out of the hole or down through the hole with the use of invisible wires. These birds did not copy the direction of plug removal.

NOT FEATHER-BRAINED
The corvid family, including
crows, ravens and magpies, have
larger brains than one would
expect for an average bird of
their body size.

See also
Captive crows show talent for tool use,
  page 20
Monkey see, monkey do?, *page 90*

THE GREAT APE DEBATE

IT IS SO WIDELY ASSUMED THAT GREAT APES HABITUALLY COPY THE ACTIONS OF OTHERS THAT "APING" HAS BECOME A SYNONYM FOR IMITATION. BUT WHILE THERE ARE MANY ANECDOTES OF APES IMITATING EACH OTHER AND HUMANS, IT IS ONLY RECENTLY THAT CONVINCING SCIENTIFIC EVIDENCE OF GREAT APE IMITATION HAS ACTUALLY EMERGED.

NATURAL HABITAT
Chimps: Western and Central Africa

One of the first influential scientific studies of ape imitation involved orang-utans at Camp Leaky in Indonesia. The orang-utans had been rescued from the pet trade and were being rehabilitated for reintroduction to the wild. Many of them apparently imitated the activities of the humans living at the camp. For example, one adult female, Siswoyo, copied weeding, cutting the weeds off at their base with a stick and then piling them up in a straight line along the path behind her, just as she'd seen the humans do. Another female, Supinah, copied some elements of fire-making by piling up sticks, pouring out fuel and even waving a lid back and forth just like the camp staff.

A "Bronx cheer"

One of the first experiments on imitation using a home-raised chimp involved Vicki. The researchers taught her a "Simon says"-like game. They moulded and shaped her responses using food rewards so that she would reproduce certain actions on the command, "Do this". After Vicki learned to match over a dozen actions, the researchers began to demonstrate actions to see if she would copy them immediately without being rewarded. They demonstrated spinning on one foot, "blowing a raspberry" (or "Bronx cheer") and touching one's nose. One action that Vicki found very difficult to copy was closing and then opening her eyes. Initially she would screw up her face, but her eyes remained open. Eventually she resorted to pulling her eyelids closed with her fingers! Although Vicki's performance seems impressive, the report did not cover her behaviour in any detail and much of the time it is very difficult to work out exactly what she did. Hence the need for a further experiment,

with two young chimpanzees, Scott and Katrina, by different researchers. The chimps imitated about one-third of the 48 novel actions shown to them but their copies were often rather crude. They copied such actions as touching the backs of their heads, touching their noses, clapping the backs of their hands and opening and closing their mouths.

A later experiment used an adult male orang-utan, Chantek, who had been taught American Sign Language when young. He had much more experience of copying games, and performed better than Scott and Katrina, imitating more actions much more precisely. When other researchers heard about a female gorilla called Zura, who would copy humans just for fun, they thought their luck was in. Zura had been raised by humans in Ohio, but then came to live at the San Francisco Zoo. The experimenters showed Zura seven novel actions and, without receiving any rewards, she produced several imitations, albeit crude ones.

Chimps outperform children

One of the most interesting recent tests involved a version of the "Simon says" game, but the actions were performed on objects. The researchers used three language-trained and three non-language-trained chimps and compared them with eight 18-month-old and eight 30-month-old children. The experimenters would give a chimp one object for four minutes to see if it might spontaneously produce the action that was going to be demonstrated. If it didn't perform the action, the experimenter took back the object and said, "Do this", demonstrated the action and then handed back the object. The chimps and children were shown actions such as rubbing a scrubbing brush back

APING HUMANS

Published anecdotal accounts of great apes imitating the actions of humans date back over 100 years. This story, from 1869, was told by Menault. M. Flourens explained how one day: "I paid him [the orang-utan] a visit, accompanied by an illustrious old gentleman, who was a clever, shrewd observer. His somewhat peculiar costume, bent body and slow, feeble walk at once attracted the attention of the young animal, who, while doing most complacently all that was required of him, kept his eyes fixed on the object of his curiosity. We were about leaving, when he approached his new visitor, and, with mingled gentleness and mischief, took the stick which he carried, and pretending to lean upon it, rounding his shoulders, and slackening his pace, walked round the room, imitating the figure and gait of my old friend. He then gave him back the stick of his own accord, and we took our leave, convinced that he also knew how to observe."

This story, from 1925, concerns Robert Yerkes, founder of the Yerkes Primate Research Centre in Atlanta, Georgia, and a young pygmy chimpanzee called Chim, which he had hand-reared. "A boy of 12 who was playing with Chim in the New Hampshire pasture one day began to spit to see whether Chim would imitate him. Chim watched with keen interest and perfect attention. Almost immediately he tried to spit. His efforts were amusing if not effective. The following day in the observation room he was seen off in a corner practising spitting, having achieved in the meantime a fair degree of proficiency. As this performance was promptly discouraged the story stops here."

and forth on the floor, squeezing the bristles of the brush, opening a pressure lid with a screwdriver and striking the surface of the lid several times with the screwdriver. The language-trained chimps and children performed very well, but the non-language trained apes imitated very little. In a second experiment, the researchers demonstrated an action, but the object was not given back until 48 hours later. Not only did the children and language-trained chimpanzees prove capable of delayed imitation, but the chimps actually performed significantly better than the children.

Profitable problem solving

All the above ape imitations involve functionless acts, but a series of experiments has been conducted with chimps, orang-utans and gorillas using a puzzle box designed to be like a kind of fruit. The outer casing needed to be pared back to reach an inner kernel of food. The food was locked inside using two types of latches. One consisted of a T-shaped pin and a barrel with a lip. The pin was either pulled straight out, or rotated several times before the lipped barrel was either pulled or turned to free the lid. The second latch consisted of a pair of smooth rods that were either rotated while being pulled or poked out. The gorillas and orang-utans showed little ability to match the methods used to remove the rods, but they did tend to rotate the pin if they had seen this demonstrated. It is quite possible they learned this by emulation; that is recreating the movement of the object rather than copying the precise actions of the demonstrator. The chimps did match the twisting action while pulling the rods, though. Their copying seemed to go beyond emulation because they tended to remove the rods in either direction, regardless of the method demonstrated. They matched the actions used by the demonstrator rather than just recreating the movements of the objects. It therefore seems that the great apes can imitate relatively functionless actions, and chimps, at least, do show some ability to copy actions that help to solve a problem.

See also
Can you really teach a
   chimp to speak?,
   page 150
An orang-utan learns to
   sign, *page 164*

APE CULTURE?

NATURAL HABITAT

Orang-utans: Borneo and Sumatra

Gorillas: Western Central Africa

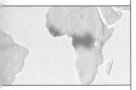

Chimpanzees: Western and Central Africa

RESEARCH HAS SHOWN THAT GREAT APES CAN IMITATE ACTIONS WITHIN THE CORRECT FUNCTIONAL CONTEXT, WHICH IS CONSIDERED BY SOME TO BE A REQUISITE FOR CULTURAL TRANSMISSION. THERE IS THEREFORE GROWING EVIDENCE THAT BOTH WILD AND CAPTIVE GREAT APES POSSESS SIMPLE CULTURES.

Human culture is a dizzyingly complex edifice and touches every aspect of our lives. Not only do we pass down problem-solving skills from one generation to the next, but we also pass on our shared history and beliefs. Without the capacity for language, animals cannot be expected to rival the scope or complexity of human culture. Nevertheless, some species, most notably the great apes, show some capacity for simple cultural transmission.

Shared behaviour

The most robust evidence for cultural transmission in a wild species comes from chimpanzees. In one major review, researchers identified 39 different behavioural patterns related to tool use, grooming and courtship that were present in some groups of wild chimpanzee but not others. For example, in a few groups of chimps, individual members like to draw leaves between their teeth to make a loud, rasping sound. One group performs this so-called leaf-clipping behaviour during courtship, while members of another group do it while warming up for a fight. The members of one group of chimps possess the strange habit of grooming leaves and, in another group, the members like to scratch each other vigorously as part of their grooming routine.

When dipping for safari ants, two neighbouring groups of chimps actually use different methods. The ants have a very nasty bite, so the chimps use sticks for dipping into the ants' nest. One group uses short sticks – they dip them for a short time and then very quickly bite the ants off the surface before dipping again. The second group uses long sticks, collecting a great ball of ants that they sweep off the surface of the stick very quickly with one hand, and evenly more quickly transfer to their mouths. The researchers carefully considered whether there might be an

CULTURED ORANG-UTANS AND GORILLAS

Although, as yet, the experimental data is absent, there exist observations of orang-utans and wild gorillas that suggest they also possess a capacity for culture. One group of researchers collated data from six different field sites and found many behavioural differences between the different groups of orang-utans. At one site the apes hunted and ate slow lorises and also ate the seeds of the neesia tree. Although other forests contained loris and neesias, the orang-utans living there did not eat them. In two forests the orang-utans masturbated using sticks, while elsewhere this practice had never been observed. And at one site the orang-utans blew a "raspberry" (or "Bronx cheer") before settling down to sleep, a ritual that was absent in the other sites.

Similar cross-site comparisons are yet to be conducted with gorillas. Nevertheless, one observer has observed complex feeding behaviours in mountain gorillas that seem to have been passed down from one generation to the next. For example, consider the way they eat giant nettles that impart a savage sting, especially to their delicate lips. The researcher noticed that the gorillas break off a stem, strip the leaves, fold the leaves over their thumb and then extract the thumb holding the leaves in a tight bundle. They then poke the stinging edges of the leaves into the centre of the bundle, pick out any debris and, finally, draw back their lips, biting through the bundle with their bared teeth. The gorillas are so consistent in the method they use that it's argued that it is unlikely that they all learned such a complicated method via independent learning.

See also
The wild chimp's toolkit,
    page 34
Use of spears by wild
    chimpanzees, *page 38*

THE PAN-PIPE

High-ranking females from two groups of chimpanzees were taught contrasting methods for operating the "pan-pipe" apparatus.

1| The group of chimpanzees who saw their female use a stick to poke a trap-door and release a food reward all learned to poke.

2| The second group who saw their female lift a catch on top of the pan-pipe all learned to lift.

3| A third group of chimpanzees who received no demonstrations failed to solve the pan-pipe task.

GOOD STUDENTS

Experiments with captive chimpanzees have shown that they can learn specific techniques for solving problems that spread faithfully within and across groups.

NATURAL IMITATORS
Great apes imitate and replicate
many patterns of behaviour such
as those related to courtship,
grooming and the use of tools.

ecological factor that could account for these differences and, although there were no obvious explanations, some have argued that it is impossible to prove that these differences were categorically due to imitative learning.

"Pan-pipes"

In order to answer the skeptics, other researchers conducted an experiment on cultural transmission in captive chimpanzees. They designed an experimental apparatus called the "pan-pipe", and there were two different techniques for getting the food inside it. You could either poke open a small flap with a stick and then push a block until the food fell onto a ramp and rolled onto a collection chute, or use the stick to lift the block and release the food. Two high-ranking females were chosen to be trained as "demonstrators", each coming from a different group of chimps. One was taught the poking technique, and the other lifting.

After they had learned their respective methods, they returned to their original groups. All but two of the 32 chimps solved the pan-pipe, and most adopted the method used by their "demonstrator". Not one chimp from a third group, which lacked a "demonstrator", solved the problem. But does this provide evidence of cultural transmission in chimpanzees? (Since all the chimps lived in very similar enclosures with identical diets, the differences in behaviour can't be explained by ecological variants.)

It has been argued that to be comparable to human culture the transmission of a behaviour must be based upon imitative learning (when one learns to do something by seeing it done). And, furthermore, that only imitation could provide sufficiently faithful and robust copying to allow stable transmission over time. Initially, the pan-pipe was shown to the chimps for 36 hours over 10 days. Two months later it was presented again and the two groups still showed a preference for the particular method used by their "demonstrator". Even though the chimps copied the two methods relatively faithfully, it is still possible that they learned to lift or poke the pan-pipe by a process called emulation rather than by imitation. Emulation involves learning primarily about the properties and movements of objects, rather than learning to copy the actions of a demonstrator. Therefore, the chimpanzees could have learned how the flap moved instead of copying the actions of the demonstrator. In an elegant extension of their pan-pipe study, the researchers presented the non-demonstration group with the apparatus again, but this time using a "ghost" control. The pan-pipe's block appeared to lift automatically. None of the chimpanzees learned to lift from this demonstration. It therefore now seems that we have strong experimental evidence that chimps possess the capacity for simple cultural transmission.

other animals also have varying degrees of a self-concept, based upon their ability to recognize their own reflection in a mirror. Starting with the initial, ingenious experiments with chimpanzees, many studies of mirror self-recognition have now been completed with orang-utans, several monkey species, gorillas, dolphins and even elephants, with surprising success in most cases.

CHIMPS USE MIRRORS LIKE WE DO – TO SEE HOW THEY LOOK

NATURAL HABITAT
Western and
Central Africa

THE ABILITY OF CHIMPANZEES TO UNDERSTAND THE EXACT NATURE OF A MIRROR, AND TO APPRECIATE THEIR REFLECTION IN IT, WAS STUDIED USING WHAT IS NOW KNOWN AS THE MARK TEST.

Let's begin with a crucial question: exactly how do children understand that they exist as individuals, with their own thoughts, desires and beliefs? Part of that discovery involves the emergence of the concept of the self, and this unfolds during a specific phase of cognitive development. One by-product of the many experiences that contribute to this awareness is the ability of young children to learn how a mirror works, and to understand that they are seeing their own reflection.

In the late 1960s a similar question occurred to the psychologist Gordon Gallup as he looked at himself in the mirror while shaving. How did he know that the person he saw staring back was really him? Furthermore, was there another species that might have similar capacities for self-recognition? It was an empirical question, and what followed was an ingenious experiment that has provoked a tremendous number of subsequent tests since the results were first published in 1970. Gallup had access to a large captive chimpanzee population, and designed a study to see if they could learn to recognize themselves in a mirror.

"Baseline" behaviour

First, each chimp was tested individually in a large cage, separate from other chimps. Observers watched and recorded the types of behaviour that each chimp exhibited during a specific period to establish what is known as a "baseline". This records the nature and number of the different types of behaviour that the chimpanzee engaged in before the mirror was introduced. By recording the same types of behaviour after the mirror was used, it was possible to compare the chimp's original baseline behaviour to any changes that occurred after the mirror was used.

When all the chimpanzees had been tested with and without a mirror they were then anaesthetized. While asleep they were marked with a vibrant red dye that was placed across the top

COMMUNITY LIFE
Chimpanzees live in highly complex social groups: being able to understand how they look from another's perspective could prove useful in their daily social manoeuvrings

of one ear, and also on the opposite brow ridge, the bony area above their eyes that corresponds to the location of our eyebrows. The dye was odourless and tasteless so that the chimps were completely unaware they had been marked. Next, they were tested individually in a separate cage with the mirror present, and all the animal's responses were recorded. The results for each phase of the experiment were remarkable.

During the three phases of observation the subjects never saw the observers, and this guaranteed that the chimps were not responding socially to the experimenters. When each chimp was first observed without a mirror present, the observers were interested in the types of behaviour that the chimps directed towards themselves. That is, how many times did each animal touch parts of the body that it couldn't see? They also recorded other activities to get an idea of the different kinds of behaviours that the chimps normally produced while in a cage by themselves.

Friend or foe?

Next, a large mirror was placed outside the cage, but close enough so that the chimp could see any images reflected in it. All the chimps that were tested responded similarly when they saw themselves in the mirror. They reacted as if a strange chimp had suddenly been introduced. Some chimps responded with aggressive behaviour towards that "other chimp", while others indicated that they wanted to play or be friendly. Eventually, these responses diminished and the chimps began to experiment

LOOKING GOOD
Most chimpanzees are fascinated by mirrors, and they have proven to be extremely effective forms of enrichment for captive apes.

THE RED DYE TEST

Just because the chimps looked as if they recognized themselves in the mirror was not enough evidence to support the idea that they actually did, so Gallup invented his red dye test to support his theories. One by one, each of the chimps was given access to the mirror outside the test cage, with the experimenters out of sight. The hidden observers watched to see how the chimps reacted when they saw the marks on their ear and brow ridge. Remember that the experimenters had a record of the types of behaviours that the chimps showed when no mirror was present, followed by observations of what they did when the mirror was available. The chimps immediately began to touch the red marks and sniff their fingers. It was clear from their responses that the chimps knew to direct their actions towards themselves. They were using the information that was only available in the mirror to touch the precise locations of the strange, red marks. Gallup's test showed that chimpanzees can learn to recognize themselves in a mirror. But precisely what that means about their concept of self still remains a contentious issue almost four decades later.

MIRROR MIRROR
A young chimp comes face-to-face with its own reflection.

1| Initially chimpanzees treat their mirror image as if it were another ape. They often appear to search behind the mirror as if they expect to find the other chimpanzee there.

2| After a few hours they begin to use the mirror to explore parts of their body that are usually out of sight, such as the underside of their tongue!

4| In the dye test, red odourless dye is surreptitiously placed on the chimpanzee's face. When the chimpanzee sees itself in the mirror, it will often reach up and touch the red patch on its own face rather than on the mirror image.

3| Here a chimpanzee uses the mirror to explore the underside of its thigh.

with other forms of behaviour, as if testing out what that other chimp might do. What they were actually learning about was mirror-contingent behaviour and how a mirror worked.

Once an individual chimp apparently understood something about a mirror's function, the observers began to see the animals open their mouths and stick out their tongues, examining the inside of the mouth. This was something the researchers had never seen before. The chimps also began investigating other parts of their bodies that could only be seen in a mirror, for example their backs and the underside of their legs. When each of the chimps reached this point in their understanding of how a mirror works, the observers knew that the chimps also understood that they were looking at an image of themselves. It was an astounding discovery for the chimps and the experimenters.

Stages of recognition

The ability of young children to recognize themselves in mirrors has interesting echoes in chimpanzee behaviour. In both cases, this recognition does not just suddenly appear out of nowhere but gradually emerges through several vital stages.

Our capacity to recognize that we have our own thoughts, needs and desires may be a specialized phenomenon that is available to only very few animals. This ability is called the Theory of Mind (ToM) by researchers who study this capacity in different animal species (a study called comparative cognition), and its emergence and development in children.

The sense of self

ToM can be measured through specially designed experimental tasks for young children because this awareness emerges between the ages of 3 and 4 years. Mirror self-recognition (MSR) emerges earlier in children, between the ages of 15 and 18 months, and is a necessary component of ToM. The earliest work that demonstrated MSR took place in 1972 and tested children from 18 to 24 months old using a sham-marking procedure (before the formal testing for MSR). The children's faces were dabbed without any actual marks being made so that, when the real marks were applied, the children would be less likely to notice. When a mirror was made available, the children were observed going through the same types of behavioural responses that had been previously documented in chimpanzees.

A second study also provided the same type of findings after testing 48 children from 6 to 24 months, including 24 boys and 24 girls. Each age group had the same number of subjects so that comparisons could easily be made. As with the chimpanzees, the investigators predicted that the children would go through a succession of steps before understanding that it was their

THE POLE AND HAT TEST

In the second test, a special vest was devised with a hat attached to a pole and configured with a wire frame that could be worn by the child. When the child put on the vest, the apparatus was behind the torso, and enabled the hat to be held above the child's head. With this configuration, the only way the hat could be seen was in a mirror. Once fitted with the vest, the child was seated in front of the mirror. Now the child had to look in the mirror at the hat and he or she indicated, either by looking directly up at the hat or reaching up towards it, the connection between the mirror image of the hat and the actual hat itself.

In the pole and hat test, the children wore special apparatus with a hat attached that could only be seen in a mirror.

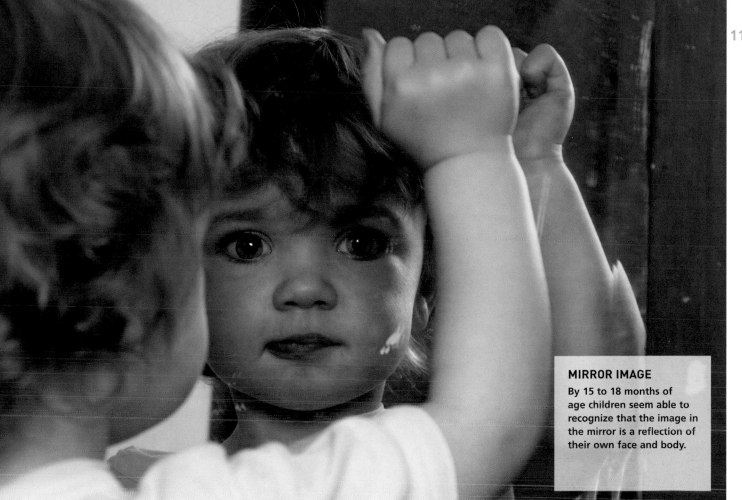

MIRROR IMAGE
By 15 to 18 months of age children seem able to recognize that the image in the mirror is a reflection of their own face and body.

own likeness that they could see in the mirror before them. The researchers also decided to explore the children's understanding of mirror images more closely by devising five additional tasks that became increasingly more challenging. In the original MSR task, the children being tested were simply marked on their faces surreptitiously with rouge, and given access to a mirror. If the children in that task – known as the Mark Test – began to examine the facial marks, the researchers concluded that the children must know that it was their own image in the mirror, and were using it to locate the marks.

Mirror testing

In the newer study, the children were first tested by putting them in front of a mirror, and they were studied to see whether they touched any place on the face or other part of their mirror image. This enabled a later comparison to be made with the formal Mark Test. The second task involved using a hat on a pole above the child's head that could only be seen in the mirror (see left). In the third, the object that appeared in the mirror was a toy that was lowered from the ceiling as the child sat in front of the mirror. The toy was not physically connected to the child as the hat had been in the previous task. Instead, it was lowered until its mirror image was just about eye level with the child. At this point, the

experimenters expected the child to see the toy in the mirror, and to turn around towards it within 30 seconds. The fourth task was the formal Mark Test in which a spot of rouge was placed on the tip of the child's nose, after which the child was placed in front of the mirror. The child's response should be either to touch the place on its face, or indicate verbally that there was something there. In the final task, the child's mother stood near the mirror, but without her image being reflected, and asked the child three times, "Who's that?" The child had to immediately respond with his or her name, or else use an appropriate personal pronoun such as, "Me!" or "That's me!"

The results confirmed the experimenters' predictions that nearly all the children responded according to their developmental age, showing the correct sequence of mirror understanding for each age group. This also confirmed the hypothesis that MSR in children does not just suddenly appear but, instead, emerges gradually through a progression of steps. In turn, the investigators concluded that the concept of the self, including whatever facet facilitates MSR in children, is not a single phenomenon. There are numerous skills and levels of conceptual understanding that are eventually integrated into a true concept of the self, most likely to occur between the ages of 3 and 4, based on the most recent findings from ToM studies.

DOLPHINS GIVE THEMSELVES ADMIRING GLANCES

DOLPHINS HAVE A REPUTATION FOR BEING INTELLIGENT, BUT CAN THEY RECOGNIZE THEMSELVES IN A MIRROR? INGENIOUS TESTS NOW PROVE THAT THEY CAN RESPOND JUST LIKE APES AND HUMANS.

NATURAL HABITAT
Worldwide, mostly in the shallower seas of the continental shelves

Dolphins are often compared with other intelligent mammals with complex and dynamic social structures and behavioural flexibility, including the great apes. A variety of cognitive approaches that have been applied to apes have therefore also been tried on dolphins, including a wide range of experiments investigating concept formation, cross-modal discrimination and gestural communication. In addition, there has been much interest in whether dolphins can recognize themselves in a mirror.

Early efforts to demonstrate dolphin mirror self-recognition (MSR) were inconclusive, but the evidence did look promising. Since that time, several new attempts have been made, and the results suggest that dolphins do show MSR. The obvious difficulty in testing dolphin MSR using a comparable approach to that used with apes is that dolphins do not have hands for touching a marked area on the body. A new experiment was necessary.

Finding the right test

A 13-year-old bottlenose dolphin was tested in two phases. In phase one, it was housed in a specially equipped pool with three reflective glass walls. During the test, a narrow plexiglass mirror was placed vertically on one wall, creating a more reflective area on that side of the pool. The second phase was conducted in two interconnecting oval pools with no reflective surfaces. However, the experimenters placed a vertical mirror just inside the gate of the smaller pool where it connected to the larger one. The pool with the three reflective sides was part of the dolphin's normal environment, where it spent part of every year. The dolphin's behaviour was videotaped for 30 minutes before, and 30 minutes after a feeding session when no mirror was present. In addition, the dolphin was either marked with non-toxic, temporary black ink marker, was pretend-marked (called "sham marking") or not touched at all.

HOW DOLPINS FUNCTION

Although most dolphin species have not yet been studied, there is a great deal of information known about bottlenose dolphins. They live in complex multi-male, multi-female groups, and form friendships and coalitions with other group members for cooperative protection and defence. Like other dolphins, they have a highly advanced system of echolocation that is used to detect objects, including potential prey, other dolphins and underwater features. The location system works when sound is produced through the top of the dolphin's head and projected forwards. It bounces off an object and reflects back to the dolphin where the feedback is translated into a "sound picture". This compensates for the remarkably poor visibility in some parts of the ocean.

DOLPHIN MSR EXPERIMENT

It has always proven difficult to test for mirror self-recognition in dolphins. Without hands, dolphins can not reach up and touch a marked part of their body. One study devised a clever sham-marking procedure to test for MSR in dolphins. A dolphin was sometimes actually marked on its body with a black marker pen, or it was touched as if it had been marked, when in fact it had not been. Subsequently, the dolphin would swim faster from one side of its pool to the other to inspect its body in a mirror after it felt itself being marked or sham marked.

The dolphin was marked in different patterns and on different areas of its body so that it would not get used to the marks, and would eventually ignore them if they were detected. These differences would also allow the experimenters to see if the dolphin oriented differently, depending on the type and location of the marks. All the videotapes were evaluated by four different individuals who were randomly assigned particular videotaped segments, and by two other "scorers" who were unaware of the experimental conditions. In addition, the actual marks on the dolphins were not discernible by the scorers, all of which should have eliminated any bias.

The researchers marked the location, duration and time of any self-directed behaviour (either mark- or sham-directed, or exploratory), non-directed behaviour, ambiguous responses and social behaviour. The mark- and sham-directed behaviour was of greatest interest since these responses occurred when the dolphin positioned itself by the reflective surface and showed orienting or repetitive body movements, which meant that it could see the mark or sham-marked areas. Exploratory behaviour included repetitive head-circling, and close viewing of the eye or genital areas in the mirror. The behaviour that was usually observed when a dolphin encountered a conspecific (that is, another dolphin) included aggressive charging or jaw-clapping or, conversely, acts including sexual posturing.

Speedy responses

The results of Phase 1 revealed that the dolphin spent more time in self-directed behaviour (inspecting parts of their body only visible in a mirror) by the reflective surface when marked, as well as more time engaged in mark-directed behaviour (touching the marks). It was also predicted that if the dolphin was really using reflective surfaces to explore its body, he would spend time by the surface with the best reflection. And, as expected, the dolphin spent more time in mark- or sham-directed behaviour

at the mirror when it was present, or at Wall 1, the best of the reflective surfaces. During Phase 2, when the mirror was placed just inside the opening to the smaller pool that adjoined a larger pool, the dolphin was stationed at the far end of the larger pool at the start of each test. Again, as predicted, it spent more time at the mirror when marked than when not marked. The researchers also predicted that once the dolphin knew of the mirror's location, it would swim faster from its starting point at the far side of the larger pool to get to the mirror when it was marked or sham-marked. This too happened. In other words, if the dolphin was using the mirror to see its own body and the marks on it (or the sham markings), it would be motivated to swim faster towards the mirror. In addition, the first behavioural response was to orientate its body towards the marked area.

Following the first dolphin's test, its male companion was also tested using the approaches that gave the clearest results with the first subject. Similar findings were observed in the second subject, and again all the predictions proved true. In conclusion, these results provide strong evidence for mirror self-recognition in dolphins, based upon data from two dolphins tested under comparable conditions. They were able to use mirrors and other reflective surfaces to investigate marked areas on their bodies, in a manner highly similar to the great apes and humans.

See also
Signature whistles in dolphins, *page 80*

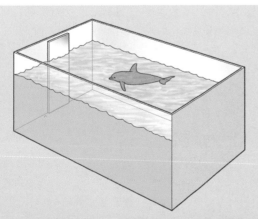

1| The dolphin was initially kept in a rectangular pool and exposed to a mirror. When it was marked, it appeared to inspect its body in the mirror, swimming back and forth in front of the mirror while turning its head and body.

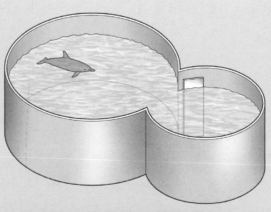

2| The dolphin was moved to a bigger pool and either marked or sham marked. After it felt itself being marked or sham marked, it would swim quickly over to the mirror – which was placed just inside the opening of a smaller pool – as if eager to check whether there was a mark.

IT'S RUDE TO STARE – GORILLAS DON'T GIVE MIRRORS A SECOND GLANCE

NATURAL HABITAT

Gorillas: Western Central Africa

Cotton-top tamarins: Northwest Colombia

CAPTIVE GORILLAS WERE TESTED FOR MIRROR SELF-RECOGNITION (MSR) FOLLOWING EXPERIMENTS WITH CHIMPANZEES, BUT DEMONSTRATED NO EVIDENCE OF RECOGNIZING THEMSELVES. RECENTLY, HOWEVER, TWO CAPTIVE GORILLAS, BOTH WITH EXTENSIVE HUMAN CONTACT, HAVE SHOWN SOME INDICATIONS OF MSR.

Following Gallup's exciting discovery in 1970 that chimpanzees could recognize their own mirror image (see page 108), it was only natural that the same experimental approach would be tried with other species. Other great ape species were the logical choice for the Mark Test, in which chimpanzees had been marked with red dye while anaesthetized, and then given access to a large mirror when awake. When the chimps examined the bright red dye on their ears and face using the reflection in the mirror, this confirmed that they understood that it was their image in the mirror. Would gorillas respond in a similar way? There was only one way to find out.

"Don't look now"

The same research group that had studied the chimps studied a small group of captive gorillas in the United States in 1981, using the same techniques, but the gorillas' reactions were entirely different, mostly because there were no responses. None of the gorillas looked at the mirror, and the social responses of the chimps were completely missing. This also meant that the gorillas did not go through the same stages of understanding how a mirror works. The obvious, immediate conclusion was that the gorillas didn't show MSR using the Mark Test. But what did this mean in terms of a concept of the self? That gorillas didn't have one? These ideas did not mesh well with what was known about the social structure of gorillas or the findings on the general evolutionary pathways of the great ape family.

MONKEY MSR TESTS

One of the most puzzling findings following the results from the now-famous studies of chimpanzee MSR has been the dramatic inability of many species of monkeys to show MSR.

There have been MSR tests with monkeys in at least 10 New World species native to Central and South America, and several types of Old World monkeys living in Africa or Asia. Among the New World kind, the capuchin and spider monkeys failed to show MSR, as did rhesus monkeys (from the macaque family), stump-, long- and lion-tailed macaques, one species of baboon (the mandrill) and, most recently, black-and-white colobus monkeys, all from the Old World. The one exception to those tested were cotton-top tamarins, a tiny South American monkey tested by dying its white topknots in brilliant neon colours. In this study, five of the six monkeys tested showed responses to the Mark Test and therefore, the author concluded, the subjects showed MSR.

ODD ONE OUT?
Gorillas are so closely related to the other animals that can recognize themselves in a mirror (humans, chimpanzees and orang-utans) that it would be very surprising if they lacked this ability.

See also
Koko, the only gorilla to learn sign language, *page 158*

However, there were clues that might explain these surprising results. First and foremost, gorillas do not make eye contact because staring is considered aggressive. That's probably why the gorillas didn't even look at the mirror. Their first glance, lasting a millisecond, just revealed another big gorilla looking back; hardly enough time to learn how a mirror works. Researchers also considered that the Mark Test was simply an inadequate or inappropriate test for gorillas, and that they were possibly the only great ape species that did not have an idea of the self.

Reliable results

More recent tests gave better results. Two gorillas – both having had close contact with humans for many years, at separate facilities – were tested to see if they could recognize themselves in a mirror. The first was the famous Koko, who had been taught American Sign Language or, more specifically, signed English, over the past 25 years and more. Koko had had extensive daily social interaction with humans, as well as opportunities to learn about virtually everything in her environment, including mirrors. When given a mirror and asked whom she saw in it, Koko often signed her name, but it remained unclear whether she actually recognized herself. It was possible that Koko had learned to give the right response from humans who pointed to the mirror in front of her and signalled her name, thereby telling her whose image it was.

Koko was now more formally tested using sham marking, with distinctive white marks being surreptitiously added to her face. When encouraged to look in the mirror, Koko immediately began to touch and investigate the marks with her fingers, indicating that she was aware of how mirrors worked, and that she knew the marks were on her face. The second captive gorilla to be tested – who had also had intensive human social interaction – responded in the same way. This suggests that gorillas do have the requisite cognitive capacity for learning how mirrors work and for recognizing their own image. It also suggests that such tests can only work on gorillas with close human contacts because they have become used to making non-threatening eye contact.

HOW GORILLAS LIVE

The gorillas' social structure in the wild is called a harem because it involves one dominant silverback male, the leader of the group, and several females with their young offspring. The silverback has to be vigilant against possible attempts by another male to steal one of his females, and against human poachers after his family.

Poaching is a big problem in Africa, with gorillas being killed for meat. Once logging became a commercial industry in many African countries, new roads were required to get the lumber out, and meat was required to feed the workers. Worse, the poachers realized that they could use the logging trucks to get gorilla meat to distant cities. The problem was exacerbated when bush meat became fashionable, and it's still served in a number of fine restaurants in East Africa. It's estimated that 6,000 to 8,000 gorillas, chimpanzees and bonobos (similar to chimps but with a narrower face, more slender frame and longer, thinner legs) are lost every year to the meat trade. A number of conservation and education efforts are underway, not least because all the great ape species are now endangered in the wild. Their numbers are rapidly dwindling, and it is likely that there will be no great apes out there, besides those in highly protected reserves, within the next 50 years.

ELEPHANTS GET THE IDEA – WITH A VERY BIG MIRROR

WHEN ELEPHANTS WERE ORIGINALLY TESTED TO SEE IF THEY COULD RECOGNIZE THEMSELVES IN A MIRROR, THE RESULTS WERE NEGATIVE. A RECENT STUDY HAS CONTRADICTED THESE FINDINGS, AND IT'S NOW KNOWN THAT ELEPHANTS – LIKE GREAT APES, CHIMPS, HUMANS AND DOLPHINS – DO HAVE THIS ABILITY.

NATURAL HABITAT
Africa, India, Sri Lanka, Southeast Asia, Malaysia, Indonesia and southern China

While the behaviour of elephants provides clear evidence of high intelligence, there has been little definitive experimental evidence. However, one capacity that has been of interest for cross-species study is mirror self-recognition (MSR), the phenomenon that emerges in children aged from 15 to 18 months whereby they understand that they are viewing their own image in a mirror.

Four signs of awareness

Confirmation of MSR has been provided for humans, all the great apes (though it applies only to gorillas who have had close contact with humans over many years, see page 116) and bottlenose dolphins (see page 114). Theses studies invariably show a four-stage process that begins with social responses towards the mirror, as if the image represents another animal. Then the mirror itself is closely inspected, with looking behind it, presumably in search of the perceived animal whose front image is visible. During the third stage the subject makes repeated behavioural responses towards the mirror, experimenting with the contingency between the subject's own body movements and the movements of its mirror image. Once there's more understanding of how the mirror functions, the subject then begins to show self-directed behaviour, using the mirror to inspect body parts that it hadn't previously been able to see.

MIRROR-GUIDED REACHING

Mirrors are not only useful for inspecting one's body, they can also be used as visual tools. Although monkeys do not seem to recognize themselves in mirrors, they will use small hand mirrors (made of polished steel) to look down hallways that would otherwise be out of sight. Elephants have shown a similar ability. They will use mirrors to locate food items that are placed out of direct sight.

1| When presented with a sufficiently large mirror, an Asian elephant will eventually begin to use it to inspect its body.

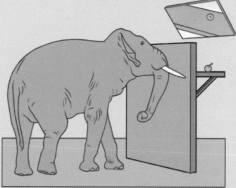

2| An Asian elephant is presented with an apple placed out of sight on a shelf on the other side of a wall. The elephant cannot directly see the fruit, but she can see its mirror image.

3| The elephant uses the mirror to reach over the wall and collect the fruit on the other side.

Until 2008, results from MSR testing of elephants did not support the idea that they were capable of recognizing themselves in a mirror. The first study of elephant MSR in 1983 was modelled on the original MSR studies of chimpanzees (see page 108). Two adult female Asian elephants in a zoo were observed for all relevant kinds of behaviour while in their indoor area. A very large mirror was then positioned in front of the indoor space – beyond the area and out of reach of their trunks – and left in place, allowing the elephants to see their reflected image 24 hours a day. For the first two days of observation they were observed for the same period of time each morning, before the zoo opened. On the third and fourth days the keepers pretended to mark the elephants on the forehead, with a large mark on the ear, the side of the head and on the foot. Both elephants were then observed and their responses recorded.

On the sixth day, formal testing for MSR began, and the mirror was removed. The elephants were now marked for real, but before this was done each one received peppermints and menthol cough drops in case the white zinc oxide ointment had even the slightest odour. The experimenters wanted to ensure that the elephants could not detect the marked areas on their bodies. After being marked, the elephants were then observed for 15 minutes, and any instances of mark-directed behaviour were recorded. The elephants were next moved to the other side of the indoor area where the mirror had been placed during the sham-marking test, and they were again closely watched and recorded. At several-day intervals after the real marking test, the animals were marked with different colours and patterns, and their responses were again recorded.

The results of the Mark Test revealed that the elephants were initially quite interested in the mirror and had responded with raised trunks; something that occurs naturally when elephants either threaten other elephants or sniff the air to detect their presence. They also inspected the mirror visually while swinging their heads back and forth, with both eyes scanning the mirror rapidly. A second form of visual inspection entailed the elephants positioning themselves so that only one eye was oriented towards the mirror. However, after that, the animals' responses to the mirror quickly dissipated. No self-directed behaviour was observed in either elephant, despite the mirror being present for two weeks. The experimenter concluded that there was no evidence for MSR.

Size matters

For almost two decades, the findings that elephants did not show MSR went uncontested until another research team attempted a similar approach, also with Asian elephants. They argued that perhaps the mirror used in the previous experiment,

ELEPHANT KINDNESS

Elephants are among the world's longest-lived species, and remain in close matriarchal and highly social groups. Females do not leave their "birth group", and remain to raise their own families under the watchful eye of their mothers, sisters, aunts, cousins and grandmothers throughout their lifetime. The group also offers guidance throughout an elephant's long period of maturation, contributing to its socialization into elephant society while helping to protect the young.

Elephants have long been recognized for various types of altruistic behaviours including coming to the aid of youngsters mired in mud and helping a calf to its feet just after birth. They also exhibit a type of unprecedented mourning behaviour, often returning to the grave of a group member to touch and explore the bones.

FAMILY GROUP

In both Asia and Africa female elephants and calves live in highly social groups in which the adults offer guidance and protection to their young.

ELEPHANT-SIZE MIRROR
Researchers using extra-large,
elephant-proof mirrors have
successfully demonstrated that
Asian elephants can recognize
their own reflection.

while large, was nonetheless too small for an elephant to view its whole body. This team of investigators also thought that the elephants needed to have physical contact with, and be able to investigate, the mirror in order to better learn about its physical and functional properties. Therefore a huge, elephant-proof mirror was constructed within reach, in the elephants' indoor area. These changes to the testing procedure made a significant difference to the elephants' responses throughout the entire experiment.

Just like the phases of MSR observed in children and the great apes, the elephants went through the four known stages for the emergence of MSR, with one elephant showing definite interest in the marks, with behaviour clearly directed at them. These exciting results now provide the critical evidence necessary to demonstrate that Asian elephants can, in fact, demonstrate that

they recognize their own image in a mirror. The implications for MSR and a concept of the self, however, continue to be debated among philosophers, psychologists and developmental and cognitive psychologists, and by practitioners of other disciplines interested in the Theory of Mind (see page 112).

See also
Why elephants use switches,
   page 32
Can elephants hear through
   their feet?, *page 86*

CHAPTER FIVE
NUMERICAL ABILITIES

Although it may not be the first thing we associate with animals, in many species sensitivity to quantity is pivotal to day-to-day survival. Through use of time, space and numbers, animals find their way to their home, nest or burrow; forage within a territory; orientate themselves or migrate over long distances; and evaluate the size of a neighbouring group. Once the memories of Clever Hans – the infamous counting horse – had faded, studies of numerical abilities in animals gained credibility and differing approaches have now revealed much about the capacity of rats, birds, lions, chimpanzees, salamanders and even ants to use measures of quantity regularly in their everyday lives.

THE CASE OF "CLEVER HANS"

HOW ONE STALLION FOOLED NEARLY EVERYONE INTO BELIEVING THAT HORSES COULD COUNT.

NATURAL HABITAT
Worldwide

It is doubtful whether there is a single article, story or book tackling numerical capacities in animals that doesn't mention Clever Hans, an Arabian stallion. In 1888, Wilhelm von Osten, a retired elementary school teacher, bought Hans after failing to teach a bear and a cat to add. While it is a mystery why von Osten would have started with these two animals, it's an even bigger puzzle why he'd then try a horse. However, according to von Osten, he not only taught the horse to count but also multiply and, later, learn the alphabet. Hans used his hoof to count out the answer and, once he had reached the correct number, he'd stop. Von Osten thought that some horses were exceptionally intelligent, and might have similar abilities as people. He really believed that he had perfected a method for teaching horses to count, no doubt based on his former teaching methods with children.

Who's deceiving whom?

Von Osten started giving demonstrations of Hans' prowess in his courtyard, and the news of the remarkable "counting horse" spread to other towns and cities throughout Germany. The press, however, increasingly accused von Osten of fraud, despite the fact that he never tried making any financial profit from the stallion. In fact it was von Osten who requested an investigation by the local school board to verify his claims about Hans' skills,

SMART HORSES
As well as having fantastic memories, horses are masters at reading and responding to – even anticipating – the most subtle human body language.

STILL CLEVER
Though Clever Hans was a fake, the more important fact is that all horses have uncanny abilities to perceive unconcious movements in humans.

and to evaluate his methods for teaching the horse to count. The board appointed a commission consisting of two teachers, a vet, a count, two zoo directors, a horse trainer, two academic scholars, two military majors, a circus manager and even a magician!

As they began their investigation, different individuals in the group were told to watch various parts of von Osten's body to see if he was secretly giving the horse clues as to when to stop counting with his hoof. But throughout the performance the investigators could find no evidence for any auditory or visual signals from von Osten. The following day their tests challenged

Hans with more difficult tasks, and involved more complicated testing arrangements. For example, von Osten had to bend forwards while standing behind one of the commission members, who was asking Hans questions, to minimize any possible opportunity of cueing the horse. And still Hans performed very well.

When the testing was completed, the commission agreed that von Osten was not intentionally deceiving the public, but how Hans was "counting" remained a puzzle. One of the academics from the commission decided that Oskar Pfungst, one of his graduate students at the University of Berlin, should try to determine precisely how Hans arrived at the correct answers. For the next three years, Pfungst used a variety of methods to evaluate Hans' understanding of counting and other skills. He observed the stallion many times while he "counted", and designed controlled experiments to allow him to change the testing conditions to eliminate any possible forms of cheating.

Pfungst the observant

Despite what the commission had reported, Pfungst had always suspected that von Osten was somehow providing cues for Hans. He first thought that von Osten was unconsciously providing some kind of nasal cues that neither the commission nor the audiences noticed, but then realized that Hans made more errors if von Osten was further away, and performed even worse when it got dark at dusk. These critical observations allowed him to formulate more rigorous tests.

Pfungst was well trained in designing and conducting scientific experiments, and knew that it was important to control the physical environment within which Hans was performing. He therefore had a tent erected in the courtyard, preventing the public from watching. In one test, he made large cards with large numbers printed on them, and used them under two different experimental conditions. In the first, the person who asked Hans the question knew the correct answer; in the second, the questioner didn't. Remember that by now Hans was apparently able to solve extremely complex arithmetic problems, such as the square root of a very large number, a calculation that could not be done mentally by the tester. Under these conditions, Hans failed miserably if the questioner did not know the correct answer to the problem. He also performed poorly when tested on his non-counting skills, for example "reading" names written on a slate. Consequently, Pfungst's experiments conclusively showed that Hans was unable to read numbers or words.

Next, he tackled the possibility that Hans could not really answer auditory counting questions, either. This test was a simple matter of von Osten whispering a number into Hans' ear, and Pfungst whispering another number into the other ear. According to von Osten, Hans should have been able to tap his hoof until he came to the right answer, that is, the sum of the two numbers – but he failed. In later tests Hans wore a blind with an additional black cloth over it, so that he could not possibly see the questioner. Again he failed.

Pfungst finally observed that whoever gave Hans a problem bent ever so slightly forward when asking the question. Then, either just before, or as, Hans reached the correct number using his hoof, the questioner unconsciously bent backwards and just slightly upwards. That's how Hans did it – very minute movements of the upper body told the horse when to stop tapping his hoof.

SCIENTIFIC TECHNIQUES AND RELIABLE RESULTS

This type of unintentional signalling is known as social cueing, and many experiments since Clever Hans, especially numerical studies with many different species, have often been criticized because such effects have ruined the results. Thanks to Pfungst, researchers are now very aware that they must incorporate strict scientific procedures into their experiments, thereby minimizing any possibility of giving social cues to the animals in a test.

It's also worth noting that wild horses in their natural environment are a prey species, and that their laterally placed eyes have a powerful biological predisposition to detect any small movement that might signal the approach of a predator. Riders, jockeys and trainers have all learned that horses will often shy at very slight movements like a small fluttering piece of paper to the side. That is why it is standard procedure for thoroughbred racehorses to wear small blinders that cup the outside of each eye, forcing them to focus straight ahead, so that they can't be distracted by the sight of other horses. Hans was clearly very alert to the slightest movement.

A NATURAL ABILITY

Clever Hans was only showcasing a natural ability in all horses when he calculated sums. Any cues from a familiar handler can cause them to react.

YOU CAN COUNT ON AN ANT TO FIND ITS WAY HOME

THERE ARE APPROXIMATELY 12,000 KNOWN SPECIES OF ANTS, AND THEY LIVE IN ONE OF THE MOST HIGHLY ORGANIZED SOCIAL STRUCTURES OF ANY KNOWN ANIMAL ON THE PLANET. FURTHERMORE, A NEWLY DISCOVERED CAPACITY HAS REVEALED THAT THESE CREATURES HAVE AN INTERNAL MECHANISM THAT WORKS LIKE A PEDOMETER, HELPING THEM TO RETURN TO THEIR NESTS AFTER FORAGING.

NATURAL HABITAT
Sahara Desert

The Sahara Desert ant recently became the subject of intensive study, with remarkable results. Living in a monotonous habitat, the question has lingered for decades – how do such ants always take the most direct route home, and walk precisely the distance they need to travel? One group of scientists wanted to understand more about how ants navigate across a desert. It was already well known that they use celestial cues to establish a general direction home, but how could they know exactly which path would take them directly to the nest, both by the shortest distance and the least amount of time? This mystery had baffled scientists for generations.

Stilts and stumps

There have been many theories attempting to explain how ants accomplish this remarkable feat. Comparisons with other insects, such as honeybees, suggested that ants might remember visual cues that go unnoticed by human observers. But this idea was disproved by experiments showing that ants can not only see in the dark, but that they have no trouble navigating home even when blindfolded. Another theory (later disproved) proposed that ants remember the duration of the march from the nest and time their way back, while what was considered one of the most outrageous (voiced in 1904) suggested that ants could actually keep track of how

many steps they had taken as they moved further and further away from the nest during foraging expeditions. The scientist who proposed this might have been mocked then, but not now.

An ingenious set of studies was completed in 2006 that tested whether ants really have an internal mechanism that works like a pedometer, so that once they finish foraging for the day, they'll know the number of steps they must take to return home. But how could anyone ever demonstrate this type of unobservable ability, especially in a tiny desert ant? One group of researchers took up the challenge with an incredibly innovative approach. They reasoned that if ants do have an internal step-counter or pedometer, any changes to their normal stride would affect where they end up when returning to the nest. To test this hypothesis, they first trained ants to walk to a feeder along a straight aluminium channel to and from the nest. Once the ants were familiar with this route, the researchers introduced some extremely creative changes.

They modified two groups of ants' legs in dramatically different ways. When the ants had reached the food source, the researchers immobilized them temporarily in a harmless wax and, with one group, they very carefully glued thin extensions onto the ants' legs, so in effect they were on stilts, thereby lengthening their legs and stride. The ants in the second group had their legs shortened by snipping off their feet and lower legs. Left with a shorter stride, more steps were now necessary to cover the distance that they normally travelled. The ants were then allowed to make their way back to their nest but this time down an aluminium channel that did not lead to their nest.

The ants with the shorter legs walked the channel but typically stopped after travelling well short of the distance of the nest from the food source. They paced back and forth, appearing to be looking for the nest. The ants on stilts overshot the distance required to reach the nest by a similar margin and they too wandered around looking for the nest.

Over time, however, both groups learned to adjust the number of steps between nest and the food. These startling results provided the first scientific evidence that the ants' stride length is the unit of measurement used by their internal pedometer. Let us hope that the scientist who first proposed the idea is smiling in that great research laboratory in the sky, even if it took more than a century to realize that he was right.

See also
Dance language of
  honeybees, *page 62*

NO PLACE LIKE HOME
Sahara Desert ants navigate to and from their nest by keeping track of the number of steps they take as they search for food.

SAHARA ANTS ARE PUT THROUGH THEIR PACES

Scientists set up a series of experiments to discover whether Sahara Desert ants, whose habitat is often monotonous and lacking in significant landmarks, found their way to and from their nests by "counting" and "remembering" the number of steps they took as they searched for food.

The scientists tested their hypothesis by altering the number of steps the ants needed to take between their nest and a food source.

1| Ants were trained to walk along a straight path from their nest to a food source. They soon became familiar with the route.

2| Once at the food source one group of ants had their legs shortened. As a consequence, this group initially stopped short of the nest, apparently unable to find it.

3| Another group of ants had their stride extended by having tiny bristles glued to their legs. This group appeared to be lost after walking further than the distance of the nest.

LIONS SHOW ROAR TALENT

AFRICAN LIONS ARE ABLE TO DETERMINE THE NUMBER OF POSSIBLE INTRUDER LIONS NEARBY, BASED ON THEIR VOCALIZATIONS. IF THE PERCEIVED NUMBER IS GREATER THAN THEIR OWN, THEY WILL NOT CONFRONT THE LARGER GROUP.

NATURAL HABITAT
Sub-Saharan Africa

With their powerful jaws and tremendous paw strength, lions are the most formidable hunters in Africa. They can also see at night thanks to specialized tissue lining the retina that reflects light back through the retina. Like wolves and African wild dogs, they are social carnivores and live in groups called prides, and are cooperative social hunters that live on open plains with permanent water sources and abundant game. This type of social structure is very different from that of big cats that are solitary hunters. It also means that lions are quick to defend the group's territory.

Posturing and fighting

Recent research has uncovered a remarkable ability among lions to evaluate the potential outcome, should they have an aggressive encounter with other lions. When it comes to actual fighting between many animal species and conspecifics (the same species as themselves), it does not make sense to cause serious physical harm if that can be avoided. Mechanisms that allow for displays, bluffing and other

THE SOCIAL UNIT
Prides can be large or small, but one important characteristic is that all the females are related. It is extremely rare for unrelated females to form a pride.

See also
Working together – is it teamwork or just a bunch of animals?, *page 180*

THE IMPORTANCE OF THE FEMALE

While there is a great deal of difference in size and body between male and female lions, with the males much larger and instantly recognizable by the dramatic mane, the females are pivotal to the pride. They are the hunters, they defend the pride's territory and they raise the cubs. They are also capable of living on their own, but maintain close ties with their sisters and other relatives as a group. A typical pride numbers around 15 individuals, but groups may be much larger and include up to 18 females, a coalition of adult males that may or may not be related and the group's cubs.

Several females may give birth at the same time, allowing for communal litters that can be cared for by different females. This arrangement has significant advantages, both for the cubs and their mothers, and it greatly increases the chances that more of the cubs will survive.

EXTENDED FAMILY

Lionesses form the stable core of their pride. Males are often supplanted by rival males, while females remain within their birth group for their whole lives. This means that female pride members tend to be closely related to one another.

INTIMIDATING MALE

The full, preferably dark, mane of the male is one of his greatest contributions to his pride. With it he frightens any intruders, and confirms his virility to pride females.

non-aggressive, non-physical challenges are invaluable because they help minimize serious wounding and possible lethal attacks among and between lion groups. All we need to establish is whether individual lions can judge the number of individuals in nearby groups to determine the best possible course of action.

Researchers in Tanzania, East Africa, therefore tested a large number of prides living in the Serengeti National Park in an ingenious way and discovered that lions can, in fact, make a numerical assessment of another pride, adjusting their own behavioural response based on whether or not they are outnumbered. The type of acoustic features in a female lion's roar include what are described as soft, introductory moans, followed by a series of loud roars ending with several grunts in a sequence. The entire bout lasts under one minute. One female starts, and is then followed by the other females with their roars overlapping as each bout begins and ends at different times.

Since animals in one pride can recognize the roars of unfamiliar animals, this suggests that each lion has some distinguishing features of its own vocalizations that allow for individual recognition. This, in turn, indicates that it's possible to identify the number of lions roaring simultaneously. Recordings of unfamiliar roaring lions were therefore played through loudspeakers near female lion groups of varying sizes to see how they reacted.

The number of potential intruders, indicated by whether the prides heard one or three roars through the loudspeakers, had a clear effect on how the groups responded. If the group heard one lion, it would approach. If it heard three lions, it would often not approach the area where the loudspeaker was positioned. Also, if it heard three roars and did approach, the group was much more cautious, looked around at their own group members much more, and also stopped more often as they approached. All this indicates a reticence to take on more than one individual and, according to other tests, to take on a group that outnumbers theirs.

There was also some indication from the group's roaring that they may have been signalling other group members, either alerting them to the presence of intruders or calling for assistance. In fact, in 43 per cent of cases other females joined the initial group after hearing their roars. Roaring therefore served two purposes: first, communicating to potential intruders the numbers they were up against and, second, summoning help.

LIONS COUNT ROARS

Female lions roar in a distinctive manner. Other lions can recognize the roars of familiar individuals. Therefore, if unfamiliar lions roar within the territory of a pride, they are immediately alerted to a potential threat. A clever experiment was conducted to investigate whether lions can estimate the number of their rivals from sound recordings of roaring.

1| When the recording of a single lion was played, a lone listening lion would often approach the sound in a confident manner.

2| If a single lion heard a recording of three lions roaring, she would often retreat.

3| When a group of lions heard the recording of three lions roaring, they seemed cautious.

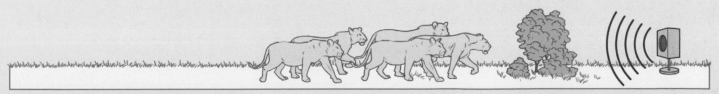

4| If the group approached the place where they could hear multiple lions roaring, they would do so in a very hesitant manner.

BIRDS: THE NUMBER CRUNCHERS

ALTHOUGH FEW STUDIES OF COUNTING IN ANIMALS WERE UNDERTAKEN AFTER CLEVER HANS, AN EXTENSIVE SERIES USING CAPTIVE BIRDS WAS COMPLETED IN THE 1940S AND STILL REPRESENTS SOME OF THE MOST RIGOROUS AND WELL-CONTROLLED ANIMAL COUNTING EXPERIMENTS.

NATURAL HABITAT

Corvids: Worldwide except for the tip of South America and the polar ice caps

Parrots: Most warm and tropical regions

With the unmasking of Clever Hans in 1907, the horse that we now know could not really count, few scientists ventured into studies of counting in non-human animals. One exception, however, stands out because of its scientific rigour and breadth, and the types of species that were studied. They raise the question, is the term "birdbrain" really justified?

"Cor, stone the crows"

In the 1940s and 50s, the researcher Otto Koehler began an extensive series of experiments to determine the extent to which several types of birds could utilize numerical information. The range of tasks that magpies, ravens, crows and other members of the corvid family were trained to complete was considerable, and has yet to be duplicated by other scientists.

The corvids appear to be the most intelligent birds, certainly in fiction. One classic tale concerns a hunter who tried to shoot a crow, but the bird kept eluding him by keeping well away. Not to be outdone by a crow, the hunter asked four of his friends to join him in a blind where the group wouldn't be seen by the bird. Eventually, each of the hunters grew tired of waiting for the crow to return, and it was only when all five men had left that the crow reappeared. True or not, this highlights the idea that some birds might be able to keep track of quantities. However, the question of why avian species would have benefited in terms of survival from such capacities is another question altogether. (As is the suggestion that birds that do not migrate may be more intelligent because they have to be more resourceful, and therefore smarter, in order to survive winter.)

Jakob

Koehler happened to have a pet raven named Jakob, and began studying the bird's ability to use numbers. In one of the tests he presented the raven with a set of five boxes, each with between two and six black dots on the lid. So when presented with, say, three of the boxes only, the bird would tap on the box with three dots. Each time the bird got the correct answer, it was given food as a reward, and eventually became highly successful at the task. One very important feature of the experiment was that the size, shape and locations of the dots on the boxes was changed in every trial. This forced the raven to pay attention only to the number of markers, and this feature of the study was probably the pivotal one for ensuring that quantity alone was critical.

The scientist then gave a parakeet the same task and found that, after training to work up to the larger numbers, it could also correctly match the dots and objects. Next, a parrot was given a similar test. Food was put in boxes, and they were lined up in a row but the parrot was permitted to open the ninth box only. It was

BIG BRAIN

The corvid family has the largest overall brain size of all the birds. The crow, for example, has the same relative brain size as the chimpanzee in terms of its brain/body ratio. Both species are also highly social, and it has been suggested that the evolutionary pressure for greater intelligence in early humans was fuelled by the demands of living in a dynamic social structure where cooperation, individual recognition and an understanding of complex social interactions was imperative. Extending such social processes to corvids is not difficult.

Also note that birds' brains are uniquely organized, and differ from a chimpanzee's (and a human's) that depend upon the extensive cortex that covers the entire top of the brain and allows for flexible behaviour, learning, and the entire gamut of higher order information processing. Birds completely lack a cortex and instead have a portion of brain called the hyperstriatum that allows their brains to perform a variety of functions. Ravens, crows and magpies have the largest hyperstriata of all the birds, and the largest overall brains with a huge number of brain cells.

See also
Birdbrained is best, *page 96*

observed bowing its head at each box until it came to the ninth, whereupon it received the food inside the box. Remarkably, during one test session, someone knocked on the door apparently disrupting the bird's count, whereupon the parrot went back to the beginning and started all over again!

Unfortunately, these earlier studies did not have much impact on other studies of animal counting at the time because they were published in German and were not widely accessible by English-speaking researchers. They've since been translated and recognized for their significant contribution towards our understanding of other species' capacity for understanding quantity.

BRAINY BIRDS
There are over 120 species of corvids, including ravens, rooks, crows, jackdaws, magpies, jays and nutcrackers.

RATS THAT CAN COUNT

NATURAL HABITAT
Worldwide

RATS CAN MAKE NUMERICAL JUDGEMENTS ABOUT THE SPECIFIC POSITIONING OF AN OBJECT, USE A COUNTING MECHANISM TO PREDICT WHAT WILL HAPPEN NEXT, AND RELY ON AN INTERNAL TIMER.

Rats, as a species, have thrived in a wide range of environments and live within a complex social organization, both of which may reflect notions of a flexible intelligence. Recent laboratory studies have documented that, among other skills, rats have a concept of numbers.

A sharp memory

One of the simpler tasks involved a long, straight runway or alley with a small amount of food at the very end – the "goal box". The rats were taught to run down the alley and find food in the first two or three trials, but not in the last of the series. Two possible fixed sequences were randomly used, and consisted of one sequence that had a non-rewarded trial (N) and two rewarded trials (RR) followed by a non-rewarded trial (N), and a second sequence that included two rewarded trials (RR) following by a non-rewarded one (N). This meant that the trials varied, and sometimes included two trials with a food reward and none on the third, or one with no food, then two with food, with no reward on the fourth.

RATS AND SCIENCE

This is an example of the species known as the Norway rat. Many generations have been bred for research, including albino rats that have limited eyesight, and the black-and-white version, when the experiment is designed for a rat that has good vision.

Once the animals had learned run to down the alley, they were timed in each trial during the tests. The rats ran very fast in the trials when food was expected, but very slowly when they anticipated no reward. The scientists concluded that the rats must have been keeping track of or counting the trials to react in this way. In another clever experiment by the same researchers, they discovered that rats could keep track of the types of rewards they received, such as fruit-flavoured cereal or less palatable dry pellets, after a series of fixed-order trials. After several sessions, the rats ran faster when they expected a tasty reward. These results showed that the animals were keeping track of the trials and the two possible, different rewards.

Boxing clever

Rats have also shown the capacity to learn which box in a series of identical boxes has a reward hidden inside, based on its numerical position. For example, a rat might have to learn that the correct box is the fourth in an array of six or even 12 boxes, arranged equidistantly in a line. In one study, the rats were correct even when the correct box was 10th in a line of 18. To make the test more complicated, the experimenters changed the number of boxes in every trial, but the animals always picked the correct one, never getting confused. Even when small and large boxes were used to ensure that the rats were not using the cumulative length of the boxes as a clue, they solved the problem.

TIME OUT

Rats are also expert at learning about time intervals. Studies have been conducted when rats had to press a bar in their test box after a specific period of time elapsed, say, 15 seconds, to receive a small reward of food. If the rat pressed too early, the clock reset and the rat had to wait another 15 seconds before pressing the bar again. Recent experiments suggest that the same mental mechanism supporting rats' numerical abilities can also function flexibly as a timer, when necessary. That is, rats can use a neural processor in their brains as a timer or counter if the task demands either timing or counting.

FLEXIBLE FEEDERS

In their natural habitat, rats are adaptable and opportunistic foragers. They have many predators, so it is important that they navigate their environment in the most efficient and flexible manner possible.

THE EVIDENCE

In this experiment, the rat enters a small chamber, and is faced with six boxes, each with a hinged door that is closed. An odourless edible treat is concealed behind one of the doors.

1| The rat enters the chamber, sniffs around and, after 20 seconds opens door no. 2, which is empty. The rat is taken out.

2| The same rat re-enters the chamber. On the second attempt, the rat explores further for a full 42 seconds before opening door no. 6 – another empty box.

3| The same rat re-enters the same chamber for a third attempt. After another period of exploration, lasting 40 seconds the rat opens door no. 4. Success!

4| The rat is returned to the chamber to test its knowledge. The rat now appears to have learned where the food is and heads straight for door no. 4.

UNIQUE CREATURE

The red-backed salamander
lives on land its whole life, and
breathes directly through its skin.

SALAMANDERS PREFER MORE

NATURAL HABITAT
Eastern North America

TIME, SPACE AND NUMBERS ARE INTEGRAL
FEATURES OF THE LIVES OF MANY DIFFERENT
ANIMAL SPECIES, AND AWARENESS OF QUANTITY
HAS EVEN BEEN FOUND IN AMPHIBIANS.

Many species of mammals and birds have been studied for their
ability to show some understanding of numerical competence,
typically under controlled laboratory conditions. It may be that
this ability reflects evolutionary adaptations for use within a
species' natural environment that has somehow contributed to
the species' success. One intriguing test, which has been tried on
several primate species – including squirrels, capuchin and rhesus
macaque monkeys, chimpanzees, orang-utans and gorillas –
required the animals to choose between two plates of food, one
of which contained more food than the other. However, when
the subject chose the dish with the larger amount, it received
the smaller amount as a reward. In other words, the animals
were supposed to learn that they would actually receive the
rejected plate.

Bigger is better

Surprisingly, most animals tested under these conditions were
unable to understand the rules. They persisted in choosing the
larger of the two amounts, even though they repeatedly received

CHOICE OF QUANTITIES

In another experiment, the salamanders had a
choice of four and six flies, with the six being
eventually chosen. Remarkably, these are the
same results achieved in a similar task using
rhesus monkeys and human infants. However,
it seems that there may be some limit on the
ability of the salamanders to discriminate.
In the final experiment, the salamanders
could choose between one or two flies, and
reliably chose the tube with two. Overall, the
study showed that the limit for a salamander
to discriminate between quantities of fruit
flies is three. Moreover, the experimenters
demonstrated that, like other species tested
for their ability to choose between two
quantities, salamanders always go for the
biggest quantity.

SMART CUSTOMER

It's easy to think that an animal
as apparently primitive as a
salamander responds to its prey
in a rather automatic stereotyped
manner. However, salamanders
have proven more discerning.
They can respond to quantity.
If they are given a choice
of approaching two tubes
containing flies, they prefer
the tube that contains the
greater quantity.

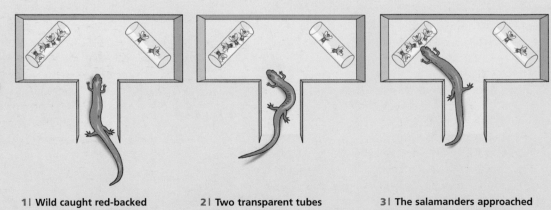

1| Wild caught red-backed
salamanders were tested in
a T-shaped chamber.

2| Two transparent tubes
containing different numbers
of fruit flies were placed at
opposite ends of the chamber.

3| The salamanders approached
the tubes that contained the
larger number of flies, even
when the difference was as
small as four versus two.

See also
You can count on an ant to find
its way home, *page 130*

the smaller reward. Some researchers believed that the mere presence of food in front of the animals made it too difficult for them to reject the largest amount. Others thought that the subjects were drawn perceptually to the amount that looked like the largest, even if sometimes the dishes contained inedible items such as rocks. At other times, the plates contained both small and large items so that it looked as if they had similar amounts, when in fact they either had exactly the same, or one plate contained more food. It seems that larger items are always chosen, even when the rejected plate had a group of smaller items whose total actually outweighed the one or two larger items in the selected dish. To many animals, then, more is better and, to an even greater extent, so is bigger.

Most of the studies tackling quantity comparisons had been done with primates. Because of this, one hypothesis suggested that the results were influenced by the optimal strategies that the mammals used for foraging. After all, a nearby tree with an abundance of ripe fruit is certainly a better feeding site than a tree that's further away with unripe fruit. But what about other types of animals besides foraging primates? This question intrigued one scientist who decided to test the idea with a non-mammal species, the salamander.

Nosing ahead

The red-backed salamander was chosen, and a number of these small amphibians were collected from a forest in the state of Virginia. The experimenters housed each salamander in its own separate box, and used fruit flies as live prey that the salamanders could spot flying around. The test had previously been tried on several types of monkeys, chimps and even human infants but, as you might imagine, the testing arrangements for the salamanders had to be very different.

Because the salamanders had been taken from their natural habitat, the scientists had to be sensitive to their lifestyle. This meant that the salamanders had to be housed and tested alone because they lived alone, and close proximity to each other during testing might result in aggression. The first experiment used a male salamander that was placed in a small, plastic box and fed five fruit flies. Next, the box was placed in a T-shaped clear plastic testing chamber. At each end of the chamber, a transparent plastic tube was inserted. The salamander was released and allowed to explore the testing area, and these procedures were repeated the next day. On the fourth day, the salamander was confined to its box by a small door, preventing it from entering the test chamber. At that point, the experimenters replaced the two plastic tubes with two new plastic tubes containing different numbers of fruit flies. Though they were confined to the tubes, there was still enough room for the flies to move around inside them. The door to the test chamber was removed, and the salamander was free to enter the area.

The researchers were interested in determining two things; first, how long would it take the salamander to choose one of the tubes and, second, would the subjects prefer the tube that contained more flies? If the salamander touched one of the tubes with its nose, that was considered a response, and the time taken to make the choice was noted. Initially, there were only two and three flies in the different tubes but, nevertheless, the salamander preferred the tube with three. When a second group of all female salamanders was tested, the results were exactly the same.

YOUNG CHIMPS LEARN TO COUNT

CHIMPS HAVE BEEN TRAINED IN ONE-TO-ONE RELATIONSHIPS AND TO MATCH THE CORRECT ARABIC NUMERAL TO A SPECIFIC NUMBER OF ITEMS BUT, QUITE SURPRISINGLY, THEY EVENTUALLY WENT ONE STEP FURTHER AND SHOWED AN ABILITY TO COUNT, WITH NO PRIOR TRAINING.

NATURAL HABITAT
Western and Central Africa

Among the animals that have been studied for their capacity to understand numbers, chimpanzees reign supreme. Studies over the past decade have identified a range of abilities showing that chimps can learn to use number symbols and correctly apply them to collections of objects, rapidly estimate the number of dots that appear very briefly, and put numbers in an ordered sequence more rapidly than college students! All in all, chimps have demonstrated a far more complex processing ability, when using numbers, than other animals tested so far. They can also choose the larger of two sets of combined objects (when compared with another pair). Some experts have suggested that the necessary intellectual abilities required to live in a cooperative and highly dynamic social structure may require quantitative skills for keeping track of hierarchical relationships among all the group members. Perhaps this applied to our early human ancestors, as it still does to modern-day chimps.

Teaching chimps

To start at the beginning, how do you teach number skills to chimps? In one key test, researchers started with chimps about 3½ years old. They were first introduced to a very simple game that was supposed to help them learn about one-to-one correspondence. They learned to count a group of objects, with one count word being assigned to one item in the group, and needed to understand the one-to-one relationship between the actual counting act, the number and the individual object being counted. But how do you communicate this conceptually to a non-verbal, non-human species?

The teacher used a very simple approach by using a bowl of small, wooden spools and an ice cube tray. The tray had 12 separate compartments that would ordinarily have been filled with water and placed in a freezer. To teach the chimps one-to-

NUMBER-CRUNCHING CHIMPS

During the 1950s, there was an attempt to teach a young chimpanzee named Viki to talk (see page 150). Although Viki learned to say only four words with great difficulty, the husband-and-wife team of psychologists who raised her also tested her on a variety of concepts, including trying to teach her about numbers. They wanted Viki to match the number of dots on a card, but Viki found it increasingly difficult as the number of dots was increased, despite the fact that there were fewer than five. In frustration, Viki began to tear up the cards whenever her teachers wanted her to start "counting". Clearly, matching dots on cards was of little interest to an active young chimp.

Another early attempt with chimps, this time under strict laboratory conditions, saw two youngsters trying to earn food by using a series of levers that turned on a series of lights. The trickiest part of their number training was that the system was based on a binary code, and not decimal-based numbers. While the teacher admitted that the chimps were not really "counting" in any manner similar to humans, it was clear that the chimps had learned a very complex system, and were able to respond accurately after many thousands of trials.

See also
Can you really teach a chimp
to speak?, *page 150*

WHO'S COUNTING?
Scientists believe that keeping
track of the social status of
group members in complex
and hierarchical social structure
may be at the root of chimps'
skill with numbers.

one correspondence, they were shown how to put one spool in each of the 12 compartments. Since only one spool would fit, the tray was perfect for the task.

Just one chimp worked at a time, and initially was given one spool. When it put the spool in the tray, it received a small treat, such as a raisin or a slice of banana. Next, the teacher gave the chimp two or three spools to place in the tray, again followed by a reward. Eventually each chimp had to put one spool in each of the compartments before getting a treat. When all three chimps were doing well, they were given a completely new test, even though their teacher had no real way of knowing if they had actually grasped the idea of one-to-one correspondence. Three stages of tests followed using gumdrops and magnets, Arabic numerals and hidden oranges (see page 146). The tests confirmed that chimps can not only be trained in one-to-one relationships but that they have an ability to count beyond what they were taught.

CHILDREN AND CHIMPS

Chimps were tested to see if they understood numbers by being shown oranges hidden around their teaching area and asked to keep track of the exact number of oranges they were shown (see page 147). The interesting point about these tests with oranges is that they show similarities between chimps and young children. Children as young as 3 spontaneously invent addition-type algorithms (rules for solving a mathematical problem in a finite number of steps) that allow them to make the same type of judgements demonstrated by the chimps. From simple counting games that children learn and practise, come remarkable skills that allow them to acquire an understanding of increasing and decreasing numbers, among other things. The precise mechanisms and processes that support such feats remain unknown in chimps and humans, but suggest that both have shared a long evolutionary history.

STAGE 1: GUMDROPS AND MAGNETS

Three yellow, plastic, circular lids were lined up in a row in front of the chimps. Above the three lids was a small tray. The teacher glued a round, black magnet to one of the lids, and left the other two empty. Next, she put a gumdrop on the tray. As the chimp watched, the teacher picked up the gumdrop, and placed it on top of the magnet while saying, "Look. One!" The chimp was then allowed to eat the gumdrop. Next, the teacher moved the lid to a different position in the row, and again put a gumdrop on the tray. This time the teacher waited until the chimp picked one of the three lids. If it was the lid with the one magnet, the teacher picked up the sweet, placed it on top, and once again said, "One!" and gave the chimp a reward.

Then the rules changed. The teacher put two gumdrops on the tray, and added two magnets to one of the empty lids. Now the chimps had to ignore the lid with only one magnet, and track the lid with two magnets. Gradually, the chimps learned to match the two magnets to the tray with two gumdrops.

Next, their teacher presented them with trays on which there might be one gumdrop or two. Now they had to scrutinize how many sweets were on the tray since it might be one or two. Again the teacher moved the position of the lids and the chimps had to follow them carefully. In time, the chimps were able to match the number of sweets to the correct lid.

In the final phase, the teacher added three magnets to the last, empty lid and used from one to three gumdrops, while also moving the position of the lids. The introduction of three magnets initially confused the chimps, but they were eventually successful in matching it to three gumdrops. But were the chimps actually counting? Perhaps they were simply matching the number of sweets with the pattern of the magnets on each lid. Maybe the chimps had learned a pattern-matching task and were not using numbers at all. Only further testing will answer that question.

1| First, the chimps were presented with just one gumdrop on the tray and three lids, one of which had a single magnet glued to it. The chimps then had to select a lid.

2| The chimps were then given the choice of matching either one or two gumdrops to a lid with one magnet, a lid with two magnets or an empty lid.

3| Finally, the chimps were shown three lids with one, two or three magnets glued to them to choose from, with the possibility of being presented with one, two or three gumdrops.

4| During the course of the experiment, the chimps chose the lid they thought was the correct match by pointing at it. If they were correct they were then rewarded with the gumdrops.

STAGE 2: ARABIC NUMERALS

The second stage involved teaching the chimps to use Arabic numerals to represent quantities. They had to understand the association between the number of sweets being used and specific symbols. To introduce number symbols, the lid with only one magnet glued on top was taken away, and was replaced by a rectangular, clear plastic placard with a large black number 1 on it. Now their choices included two yellow lids – one with two magnets, the other with three and the placard with the Arabic numeral.

1| With the placard in place instead of a single magnet, a single gumdrop was placed on the tray. The chimps chose the new placard with the numeral 1. It is likely that they began using the numeral by exclusion because they knew they should not pick either of the other lids.

2| The lid with two magnets was replaced by a plastic placard with the figure 2. It took the chimps some time to sort out the new relationships. But, eventually, they did pick the numeral that corresponded with the number of gumdrops on the tray.

3| The lid with three magnets was replaced with a placard with the numeral 3 on it. Now all three of the choices had Arabic numerals. The chimps had already used the numerals 1 and 2, so when 3 was introduced, and three gumdrops were presented they selected the newest placard with the numeral 3 on it.

4| Now, the teacher introduced the numeral 0, but she did not put any sweets on the tray and because she didn't want to confuse the chimps by giving them a reward for selecting zero she gave them a kiss on the cheek instead! The chimps understood the notion of zero, and then in a short space of time, the number 4.

STAGE 3: THE ORANGE TEST

Researchers had to check that the chimps really did understand numbers so they designed a third test. The chimps were accustomed to working with their teacher on an elevated wooden platform. Three hiding sites were established within the chimps' regular teaching space, so that the contents of each could not be seen from the other sites or from the platform. The researchers hid up to four oranges (but sometimes none) in the three sites. The chimp was shown the oranges and required to keep track of how many she saw.

Next, she had to return to the platform and select the number representing the total number of oranges that she had seen.

For the first two trials, the teacher went around the room with the chimp, pointing out the oranges as they passed each hiding place. When the pair got back to the platform, the teacher asked, "How many oranges did you see?" While chimps that interact closely with humans understand a great deal of spoken English, it is unlikely that they'd understand the entire sentence but the word "oranges" would be familiar.

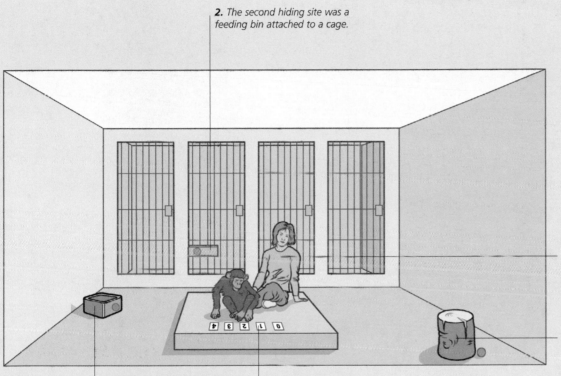

2. *The second hiding site was a feeding bin attached to a cage.*

One researcher was always present on the platform when the chimp selected a number symbol.

1. *The first hiding site was a small tree stump at the far end of the teaching area.*

3. *The third hiding site was a plastic tub 3 m (10 ft) from the elevated wooden platform.*

The chimps worked with a researcher on an elevated wooden platform. This is where the number placards were placed.

On the first trial, the chimp selected the correct number. The same thing occurred on the second trial.

On the third trial, and for every trial thereafter, the teacher stayed on the platform and the chimp went around by herself. Again she correctly selected the number symbol that represented the total number of oranges that had been hidden in the three sites. In the initial training session, the chimpanzee was 80 per cent correct.

Finally, two researchers were used. One hid the oranges and then went off, and the second sat next to the chimp when she gave her answer, but crucially did not know how many oranges had been hidden. This is called a "double-blind" test, and it guarantees that the researcher can't inadvertently prompt the chimp to give the correct answer. As before, the chimp got 80 per cent right and the researchers were astounded. Through some process that remains unknown, the chimp had come to understand something more about numbers than had been explicitly taught. Even more astounding was the fact that when a second phase of the experiment was introduced, and Arabic numerals were used in place of oranges, the chimpanzee again responded correctly right from the beginning of testing.

CHAPTER SIX
ANIMAL LANGUAGE STUDIES

Some of the most creative and pioneering studies completed in the last half-century include those that explore animal capacity for acquisition of language. Using approaches including American Sign Language, abstract symbols and plastic shapes that stood for words, chimps, dolphins, sea lions, an orang-utan and a gorilla have all proved to be adept language students. These accomplishments, embraced by the general public, brought about great disagreement among academics, who insisted that language was reserved solely for humans. While most now accept that some animals are capable of using and understanding abstract symbols, questions about the uniqueness of human language remain unanswered.

CAN YOU REALLY TEACH A CHIMP TO SPEAK?

IN THE 20TH CENTURY, AN ATTEMPT WAS MADE BY THE HUSBAND-AND-WIFE TEAM OF PSYCHOLOGISTS, KEITH AND CATHY HAYES, TO TEACH VIKI, A CHIMPANZEE, TO PRODUCE HUMAN SPEECH.

NATURAL HABITAT
Western and Central Africa

Chimpanzees share a significant portion of their genetic material, or DNA, with humans. They overlap closely in terms of their morphology, physiology, anatomy, gestural communication and cognitive abilities. So it wasn't surprising when two psychologists, Keith and Cathy Hayes, began trying to teach a very young chimp named Viki to talk.

Initial success and fame

Despite the fact that chimps are anatomically different to humans in several key respects, Viki was able to produce two basic sounds resembling "up" and "ah". Eventually she learned to combine them to produce four word-like utterances – "mama", "papa", "cup" and "up" – that she used in the appropriate context. The first two words, "mama" and "papa", meant that she had to change the first sound with her lips in order to differentiate between the "m" and the "p". For "cup" and "up", Viki first had to produce a hard "c" sound, and again form the "p" sound with her lips. In fact all the words were difficult for Viki and, during her training, her teachers had to help her shape her lips by pressing their fingers against them. Eventually Viki learned to do this herself.

During the project, Viki became quite famous, even appearing on the cover of *Life* magazine, the most important weekly news magazine in the United States at the time. She also appeared on several television shows, and was seen by thousands of people. Unfortunately, at the young age of 6, Viki contracted encephalitis and died, leaving the Hayes, especially Cathy, grief-stricken and devastated. (Viki had made heroic efforts to learn to speak, and the Hayes let her live with them in their home rather like a child. The emotional and social ties between Viki and the Hayes were considerable; she bonded with them as if they were her parents.)

SIMILARITIES AND DIFFERENCES

Observations of captive chimpanzees have shown that they have a very expressive communication system, with a range of facial expressions and vocalizations (including laughter) that are used during greetings, play, aggressive encounters and when alarmed or fearful. Chimps also have many gestures that are used in very similar contexts as those displayed during human conversation. Consequently, the Hayes' had hoped to find a way to use Viki's flexible learning capacity and other natural, expressive abilities to train her to speak. However, they had big problems. Chimps do not have the same anatomical structure of the larynx (voicebox), throat or tongue, or the accompanying brain structures and mechanisms giving the motor control and movement necessary for speech. Undaunted, the Hayes tried to teach Viki how to control her lips and breathing in order to produce similar sounds used by humans for speech.

Although many people cite Viki's limited acquisition of speech as the definitive study showing that chimps cannot learn to talk, it is not possible to predict what she might have achieved had she lived longer (chimps can now live into their 70s). She had certainly grasped the basic concept of using spoken words communicatively with her teachers.

Given what we now know about chimpanzee cognitive capacities and their ability to acquire different types of symbol systems (eg, American Sign Language gestures, plastic shapes for words and graphic, abstract symbols), it may have been possible for Viki to learn additional spoken words to the same level of accomplishment as the language-trained apes who have recently followed in her footsteps. We certainly owe the Hayes a big debt because without their pioneering work it is unlikely that subsequent animal language projects would have been undertaken. We would have lost a vast wealth of knowledge and understanding about the great apes that has promoted many new lines of study into the evolution of human cognitive abilities.

LEARNING FROM THEM
Studying chimpanzees' cognitive abilities has given scientists insights into human cognitive abilities.

See also
Wild chimpanzees: masters of communication, *page 76*

HOW A CHIMP LEARNED SIGN LANGUAGE

NATURAL HABITAT
Western and
Central Africa

THE FIRST CHIMPANZEE TO BE TAUGHT A GESTURAL SIGN LANGUAGE WAS A YOUNG FEMALE NAMED WASHOE, UNDER THE EARLY GUIDANCE OF TWO PSYCHOLOGISTS, BEATRIX AND ALLEN GARDNER, AT THE UNIVERSITY OF NEVADA, RENO. THIS WORK WAS THEN CONTINUED – FOR THE MAJORITY OF WASHOE'S LIFE – BY DR ROGER FOUTS AND DEBORAH FOUTS AT THE UNIVERSITY OF CENTRAL WASHINGTON, ELLENSBURG.

One of the most remarkable scientific projects of the last century began in 1966, when a husband-and-wife team of psychologists acquired a 10-month-old female chimpanzee that they named Washoe (after the county where they lived, and the location of the University of Nevada, where they taught). Their aim? To be the first to test the extraordinary idea, first proposed several centuries before, that it might be possible to teach apes to use one of the sign languages learned by humans with hearing problems.

The Gardners raised Washoe in a small trailer on their property, where she had human companions looking after her around the clock. She was totally immersed in what might be called a "signing environment" because no one was allowed to speak in her presence. The only form of communication that Washoe saw being used by her teachers and companions was a version of American Sign Language, or Ameslan (ASL), for short.

V for victory

ASL also includes a gestural technique called finger spelling, enabling words for which there are no signs to be spelled out using individual characters, represented by 26 different hand positions. However, chimps lack the finger dexterity and flexibility in their wrists to make finger spelling possible, and some signs were also difficult for Washoe to make for the same reason. Consequently, what Washoe managed usually involved only one sign, or occasionally two, at a time, but not the special syntax, or word order, that is part of the full ASL language system. Nonetheless, she did begin to imitate the signs she saw being

SIGNS OF INTELLIGENCE
Chimpanzees, the closest living primate species to humans, are capable of learning and using colours and symbols to represent objects, activities and even rudimentary concepts.

See also
Chimps and humans: do
two great brains think
alike?, *page 154*

used around her, and produced them in the right context. You can imagine how exciting it must have been for the Gardners and their helpers when it became apparent that chimpanzees could learn to communicate using representational gestures.

But is it language?

While you might expect that the publication of the first results of Washoe's achievements – after several years of the project, during which she learned approximately 130 signs – would have been heralded as a crowning achievement by the scientific community, but it got a radically different response. The unprecedented discovery of a chimp's ability to learn sign language set off a fire-storm of debate that can still be resurrected among academic scholars today. Had Washoe actually learned a "language"? Or was it something else?

Scholars from a vast range of disciplines became involved, including psychologists, anthropologists, biologists, neuroscientists, speech and hearing specialists, zoologists, philosophers and, of course, linguists. Yes, the chimpanzees could use sign language gestures that stood for objects, actions and people. And certainly these capacities had never before been demonstrated in an animal, but no-one could agree on the proper definition of a language. The academic world was turned upside-down trying to define our own use of language relative to the capacities available to our closest living primate species.

In 1970, when Washoe was 5 years old, Dr Roger Fouts – who had worked extensively with Washoe while he completed his dissertation under the Gardners – moved with her to the Institute for Primate Studies at the University of Oklahoma, Norman. Fouts continued his work with the chimp, first at the University of Oklahoma – where studies of Washoe's sign language acquisition developed much further – and then (helped by his wife) at Central Washington University (CWU), Ellensburg. Meanwhile, the Gardners continued their work with four additional very young chimpanzees, anticipating that the social interaction among the chimps would result in signing abilities way beyond those of Washoe. When their project ended, the four joined Washoe and her adopted son Loulis at CWU in 1980. Washoe died in 2007 at the age of 42, leaving a legacy of unprecedented scientific achievement.

Based on the Fouts' work, as well as subsequent studies focused on teaching chimpanzees to use other language-like systems devised by humans, it is now generally agreed that, language or not, apes are clearly capable of learning and using symbols to name and request objects or activities. They also characterize conceptual relationships such as "bigger" versus "smaller", "under" or "over", "in" or "on" and "same" or "different", while using different colour names, Arabic numeral symbols, names for people and other apes, and other representational symbols that refer to their environment. While this can't be compared to the complexities of human language, the Gardners' experiment did show that a chimpanzee's mind shares so much with our own.

CHIMPS AND HUMANS: DO TWO GREAT BRAINS THINK ALIKE?

WITH THE INTENTION OF STUDYING CHIMPANZEE INTELLIGENCE, A NEW PROJECT UNDER THE DIRECTION OF DR DAVID PREMACK BEGAN IN 1967 WITH A YOUNG CHIMP NAMED SARAH. SHE WAS TAUGHT AN ARTIFICIAL LANGUAGE SYSTEM THAT WAS COMPOSED OF PLASTIC SHAPES THAT STOOD FOR WORDS, ALLOWING US TO GET A GLIMPSE OF THE SOPHISTICATED COGNITIVE ABILITIES OF AN APE LIKE NEVER BEFORE.

NATURAL HABITAT
Western and
Central Africa

While other animal language projects were carried out using sign language, Dr David Premack had another idea. Given that chimpanzees are not equipped anatomically or neurologically to produce verbal speech, he devised a unique symbol system that was composed of plastic shapes mounted on metal bases. Each shape was unique and abstract, so there was nothing about the particular shape or its colour that was in any way suggestive of what it named. For instance, the symbol for "apple" was a blue triangle, which was neither the colour nor the shape of a real apple. Nonetheless, Sarah was able to learn the associations between the symbols and their referents. Furthermore, the symbols could be mounted on a magnetized board, from the top to the bottom. So Sarah's "sentences" consisted of a small number of symbols, placed in vertical order.

By the age of 7, Sarah had a working vocabulary of 120 words, including names for her teachers, for objects (like cup and pail) in the laboratory around her and for colours, and eventually symbols that represented more complex relationships. One of the first prepositions that Sarah learned was "on". Her teachers taught her this by putting two colour symbols on top of each other, while displaying the prepositional symbol. After some training, Sarah caught on conceptually and was able to respond to her teacher's commands. Therefore, when told that "blue goes on yellow" she placed the symbol for blue on top of the yellow

one. Her accuracy with some 12 possible colour combinations was reported to be 80–90 per cent correct, an impressive performance for the first demonstrated use of rudimentary syntax by a chimpanzee. She was even able to assign the correct colour symbols to brand new foods, showing that she could apply the basic rules for labelling an object's perceptual features. Further training with Sarah introduced verbs, interrogatives, more details related to sentence structure, and numerous concepts such as "same" and "different".

Tantalizing questions

One of the most ingenious experiments devised for Sarah has given rise to a line of research in developmental psychology, and has only within the last 10 years been investigated again in non-human primates. Premack wondered if Sarah could solve a series of problems that were presented to her as videotaped scenarios, with only one possible solution. For instance, Sarah watched a human actor shiver and shift his weight as though he were very cold, while next to him was a portable heater that was unplugged, and next to that an electrical outlet. When the videotaped scene was paused, Sarah was shown several colour photographs, and they all included the same perceptual features of the video. But only one photograph depicted the "correct" answer to the problem, showing the heater plugged into the

THEORY OF MIND

Premack and his graduate student Guy Woodruff presented their ideas in a very influential paper published in 1978. The whole notion of Theory of Mind (ToM) was quickly embraced by the field of developmental psychology, and a succession of experiments with children was undertaken to determine if, and at what age, ToM emerges in humans. The literature on ToM with children is now voluminous, and it's still a hot subject with developmental psychologists. Furthermore, within the past decade a significant number of experimental studies have been conducted to investigate and demonstrate ToM more rigorously with chimpanzees, the most likely primate to have such a capacity, as the tantalizing findings with Sarah suggested. The debate – whether ToM is a specialized capacity that developed in our own cognitive evolution, or is one we share with our chimpanzee cousins – continues.

outlet so that it could be turned on to warm up the freezing young man. Sarah saw eight different scenarios, and on seven occasions chose the correct photograph.

Premack then wondered if Sarah responded as she did because she was able to put herself in the place of the actor in the video. Did she recognize that if she were in the same situation, she would want the heater to work? Sarah had seen people plug in similar heaters over winter, and knew that they then produced warm air. Or were Sarah's responses simply based on some complex associations that she had acquired through observation of such activities in the laboratory?

If Sarah did recognize that the actor faced a dilemma, and responded as she did because she empathized with him, her performance would reflect a new ability previously thought to be unique to humans. Premack referred to such a capability as the Theory of Mind, the attribution of mental states to others, so indicating that an individual recognizes that others can have beliefs, desires and knowledge different to their own. This requires the fundamental understanding that others have minds, too.

CLEVER GIRL

Sarah developed a working vocabulary of over 120 words during her work with Premack, and she also learned to use basic syntax.

See also

Yawn and the world yawns with you – empathetic responses, *page 176*

ASK A DOLPHIN AND YOU'LL GET THE RIGHT ANSWER

NATURAL HABITAT
Worldwide, mostly in the shallower seas of the continental shelves

SCIENTIFIC STUDIES OF CAPTIVE DOLPHINS HAVE BEEN EXPLORING THEIR COGNITIVE CAPACITIES. ONE OF THE FIRST SUCCESSFUL APPROACHES INVOLVED TEACHING DOLPHINS TO UNDERSTAND GESTURED COMMANDS, INCLUDING SEQUENCES OF GESTURES REQUIRING COMPLEX BEHAVIOURAL RESPONSES.

Dolphins have fascinated humans for hundreds of years, and there have been scores of stories of dolphins helping to save drowning sailors and other acts of altruism. Although the US military had apparently trained dolphins to assist in naval operations, few other experimental studies of dolphin intelligence have actually been completed. The nearest anyone seems to have gotten is training them to give a spectacular, acrobatic performance in a sea-life park, where the dolphins replicate their natural behaviour in the wild.

Herman's dolphins

Dolphin trainers often use hand gestures or visual cues, including short coloured batons, to signal the start of a routine, such as jumping high into the air and hitting the pool with an enormous splash, or leaping gracefully out of the water with other dolphins in perfect synchrony. But it was not until 1970, when the marine biologist Dr Louis Herman decided to explore the dolphin's abilities to respond to gestured signals, that he was able to uncover greater depths of the bottlenose dolphin's intellect.

Dr Herman founded the Kewalo Basin Marine Mammal Laboratory in Honolulu, Hawaii, where he was on the faculty of the University of Hawaii. The procedures for teaching the first dolphin in the project, Akeakamai, were based on basic reward principles and animal behavioural conditioning. The researchers at Kewalo Basin were interesting in seeing to what extent a dolphin could understand gestured commands from a human trainer. Akeakamai was first taught to position herself in front of her trainer, who was standing poolside. The trainer would give a specific gesture that had been arbitrarily assigned as a label to a specific object in the pool, such as a Frisbee, ball or foam float. Early in her training, Akeakamai merely had to swim to the immediate vicinity of the correct object. But once she acquired an understanding of action gestures, she might be asked to perform a more difficult task. When given the command "ball–flipper–touch" she'd swim to the ball and touch it with her flipper.

Akeakamai was an eager student, and learned the names of many objects, the names for people in her environment and symbolic reference for things that were not present. To test this, the dolphin was taught to use a set of paddles representing the words "Yes" and "No", located in different parts of the pool that she could easily reach with her beak. She then watched as her trainer tossed a number of objects into the water. When asked whether a specific object was present in the pool, Akeakamai responded correctly. So, when asked, "Is there a float?" she would touch the paddle representing Yes. When the answer was No,

Akeakamai demonstrated that she understood that the symbols represented objects that weren't in her immediate presence. And rest assured, the spectre of Clever Hans, the counting horse (see page 126) still lurks in every animal cognition laboratory. During testing, Akeakamai's trainers always wore dark eye goggles to ensure that there were no inadvertent social cues that might help her select the right answer.

COGNITIVE LEAPS

As well as understanding gestures and cues that relate to actions and objects in their environment, dolphins are able to "think" and communicate about objects that are not in their immediate presence.

See also
Sea sponges provide padded protection, *page 26*

KOKO, THE ONLY GORILLA TO LEARN SIGN LANGUAGE

IN 1972, A YOUNG FEMALE GORILLA NAMED KOKO WAS ADOPTED BY A PSYCHOLOGIST, FRANCINE PATTERSON, MARKING THE BEGINNING OF THE LONGEST PROJECT STUDYING SIGN LANGUAGE ACQUISITION BY A LOWLAND GORILLA.

NATURAL HABITAT
Western Central Africa

The mid-1970s was an exciting time in the field of psychology because three studies of symbol-use in chimpanzees were simultaneously carried out. That's when a graduate student at Stanford University in California approached the San Francisco Zoo about teaching American Sign Language (ASL) to a young gorilla, born there in 1971. Koko proved to be an excellent student, and learned three signs during the first month of the project. Patterson's study became the basis for her doctoral dissertation, and since getting her PhD she has been working with Koko, changing how the world thinks about gorillas.

Previously, there was barely any scientific literature about gorilla intelligence. Although gorillas had been observed in captivity and were being studied in the wild, there was little indication that gorillas could begin to rival the highly social and intelligent chimpanzee. For that matter, there was little to suggest that they had the potential to be highly social at all, let alone show any significant intellectual prowess. Until Koko, that is.

Privileged childhood

So what made Koko different? Koko was totally immersed in a world of sign language. The earliest signs that she made indicated that she wanted drink, food and more. When she was 3 years old, Koko moved to a new home with Patterson, and made even more progress now they were able to spend extra time together. Through her natural abilities for observational learning, it wasn't long before Koko began to produce short sentences of signs. When she was 5, Koko was thrilled to get a new gorilla companion, the 2-year-old Michael, and they learned to communicate with each other through sign language. A few years later, as they grew bigger, a new home was built for them in a wooded area, away from the city.

Mirror self-recognition

One of the most significant tests involving Koko checked to see if she was able to recognize herself in a mirror. This type of test was first reported in a classic paper on chimpanzee self-recognition in

KOKO'S SIGN LANGUAGE

Koko's current sign language vocabulary is said to exceed 1,000 words, and she uses them to ask for things and name a wide variety of toys and objects; she also has names for her keepers. Koko has apparently even invented some signs of her own. While the signs that she uses are not always the same as those used in American Sign Language (ASL), most are highly similar. Dr Patterson has said that it might be more appropriate to refer to Koko's signs as Gorilla Sign Language or Gorillaspeak.

Throughout the project, Patterson and colleague Ron Cohn documented Koko's developmental history (and Michael's, until he passed away), as well as their acquisition of sign language on videotape. In addition, a great deal of data concerning Koko's use of sign language has been collected over the years, including any signs she produced, the conversational context, how many times Koko repeated signs, and if anything out of the ordinary occurred, particularly with respect to innovative or unusual use of signs. Koko was also tested periodically to establish her understanding of her gestured vocabulary, and also has completed other types of tests, including standardized tests used to measure children's intelligence.

1970 (see page 108). Chimpanzees with previous mirror experience were marked with red dye while asleep, and when they awoke and saw themselves in a mirror, they touched the marks and investigated them using the mirror. They recognized that it was their own image, suggesting that perhaps chimpanzees have a self-concept. After publication of the study, the same author tested orang-utans and gorillas using the same approach (see page 116). The orang-utans showed the same types of behaviour as the chimps, but the gorillas simply ignored the mirror at every phase of the experiment, while monkeys have also never shown any interest in mirror self-recognition. It soon became common knowledge among primatologists that mirror self-recognition was limited to chimpanzees, orang-utans and humans. But why not gorillas?

Let's answer that by looking at Koko. She had a remarkably different upbringing from most gorillas, both in zoos and in the wild. She was raised in a highly enriched, cultural environment where she learned about all kinds of things around her that other gorillas never experience. Furthermore, when captive gorillas were first tested, zoos were not housing them in social groups, the way they live in the wild. They were typically kept in pairs, and even alone when there was too much male aggression. Consequently, there was little for the gorillas to do all day but sit and eat, and such a lifestyle didn't generate much curiosity and interest in their surroundings.

Koko was different. She was raised in a social setting, interacting with human companions, and learning that the world could be represented by gestured symbols. Koko also had time to learn how mirrors worked and, crucially, she never learned that staring among gorillas was a sign of aggression. Consequently,

when she was first given a mirror and saw another gorilla looking back at her, she was not afraid. Instead, she became curious about this new friend, and crudely learned how a mirrored surface worked. Other gorillas did not have the same opportunities and so, when faced with that image in the mirror, did what all gorillas in the wild naturally do – they immediately looked away. And that's why gorillas, except for Koko, fail the standard test for mirror self-recognition.

Recently, another gorilla that has been kept in captivity for a long time and housed alone has also shown evidence of recognizing himself in a mirror. Although he did not have quite the same kind of upbringing as Koko, exploring cultural artefacts and learning a language system that he could use symbolically, he had great affection, a stable environment and fun things to do. Like Koko, this gorilla did not grow up with other gorillas, and didn't learn that looking directly at another gorilla was asking for trouble.

See also
It's rude to stare – gorillas don't give mirrors a second glance, *page 116*

IS HE SMART, OR JUST PARROTING WHAT HE'S HEARD?

AN AFRICAN GREY PARROT WAS TAUGHT TO PRODUCE VERBAL RESPONSES TO AN ARRAY OF OBJECTS, COLOURS, MATERIALS AND QUANTITIES AFTER YEARS OF TRAINING, USING A SPECIAL APPROACH.

NATURAL HABITAT
West and Central Africa

In 1977, a unique piece of research began with Alex, a one-year-old African grey parrot named after the Avian Language Experiment. The project was the brainchild of Dr Irene Pepperberg, who was interested in the animal language work being carried out with chimpanzees, an orang-utan, dolphins, sea lions and a gorilla that were then producing remarkable results, but she wondered if other animal species were also capable of using symbols. Dr Pepperberg wanted to find out if the African grey parrot, renowned for its intelligence and long life, could be taught first to assign verbal labels to objects in its environment, and then to use those labels to "name" things under experimental test conditions. Could Alex apply verbal labels to objects when there was no possible way for him to respond correctly unless he understood, and remembered, each "word" and its referent?

Watch and copy

African grey parrots are highly social, and live, forage and interact in large flocks, all of which meant that Pepperberg could use a unique teaching approach with Alex. First, a student served as a substitute for Alex, with the parrot looking on. Pepperberg would quiz the student verbally, for example asking, "What colour?" The student would answer and, if correct, was praised. Sometimes the student would deliberately mispronounce the name, prompting

Pepperberg to respond, "Say better!" She'd then give the student another chance to respond and get it right. Then Alex had a go, and was similarly questioned. Through this "social modelling" technique, Alex saw and heard the verbal interactions between the student and teacher. The approach that Pepperberg used for over 30 years with Alex probably played a critical part in teaching him the communicative function of words.

Highly skilled

Alex's early word associations included names for things with which he liked to play. He used "hide" for the piece of rawhide that he chewed, "wood" for a wooden tongue-depressor and "nut" for cashews. His vocabulary grew and, eventually, included the names of 50 objects, the numbers one to six, five different shapes and seven colours. Alex also learned a number of terms, including "same", "different", "larger" and "smaller", which he could correctly apply. Alex could even label an object by its shape, colour and the material from which it was made. He was tested with different objects, and could respond correctly when asked if there was a difference between any two items, adding "none" if they were the same.

Overall, Alex had a vocabulary of approximately 150 words but, unlike a pet parrot that merely learns to mimic sounds and

SMART ALEX

When Alex, the subject of Dr Pepperberg's research died at the age of 31, she estimated that he had the intelligence of a 5-year-old child.

words, he appeared to understand the referent for the words he was producing. Before Pepperberg's research program with Alex, parrot intelligence had not been studied, with few scientists thinking that parrots could achieve these impressive results. In fact Pepperberg believes that Alex had the intelligence of a 5-year-old child and, in some cases, his cognitive abilities were similar to those seen in dolphins and the great apes. She also believed that Alex had not reached his cognitive potential when unexpectedly, aged 31, he was found dead in his cage – a big surprise because he was apparently in good health and African greys can live to 50 in captivity. His pathology report indicated hardening of the arteries. However, because his diet was excellent and appropriate for his species, and recent blood tests had been normal, there may have been genetic factors contributing to a heart attack or stroke.

While other birds were also being trained as part of Pepperberg's research program, Alex was the most highly trained and skilled practitioner. He was a unique individual who left a remarkable and intriguing scientific legacy.

BIRDBRAINS

Dr Irene Pepperberg's language experiments with parrots in the 1970s demonstrated that they could be trained not just to mimic but also to understand the meaning of the words they are uttering.

See also
Birdbrained is best, *page 96*

A YOUNG CHIMP KEEPS HER ANSWERS IN ORDER

NATURAL HABITAT

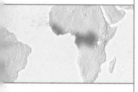

Chimps: Western and Central Africa

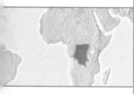

Bonobos: Democratic Republic of Congo

THE MOST FAMOUS CHIMPANZEE IN JAPAN, A FEMALE NAMED AI, WHICH MEANS "LOVE" IN JAPANESE, LEARNED TO USE AN ARTIFICIAL LANGUAGE SYSTEM ENABLING HER TO NAME, DESCRIBE AND NUMBER HOW MANY OBJECTS HAD BEEN GROUPED TOGETHER.

Ai was born in West Africa and arrived in Japan at the age of 1. When she began using a graphic symbol language, no one anticipated that she would eventually demonstrate such remarkable abilities. Ai began by learning a human-designed language that was composed of "words" represented by symbols, created using a very simple "alphabet" consisting of straight lines, open boxes, dots, circles and diagonal lines. The different elements in the alphabet could be combined to form symbols that represented objects in her environment (enabling her to name shoes, pencils and bowls, etc.), foods (including apples and bananas), playthings, people that she knew and 11 different colours (black, grey, white, red, orange, yellow, brown, green, blue, purple and pink). Now Ai had a number of symbols that could be used in new and creative ways. Soon after, the scientists who had taught Ai this language decided to test her understanding of how these symbols could be used.

Going for an A grade

First, Ai was tested on each set of the symbols that she'd been taught. Although she'd had many training sessions using object names, colours and numbers, and performed very well, she did need these symbol tests to ensure that she really understood their meaning. And whenever Ai learned new names for objects, she had to complete a test, just like children do in school, to show that she could reliably choose the correct symbol for a particular object. She also had to show that she could choose

THE FIRST COMPUTER-LITERATE APE

The LANA Project, or Language Analogue Project, founded in 1971 by Dr Duane Rumbaugh, began with a female chimp named Lana. Lana was able to use a graphic language system using a keyboard and computer. Her training allowed her to name, request and describe people, foods, objects and actions using abstract symbols called lexigrams. The lexigrams comprised a language called Yerkish, named in honor of Robert M. Yerkes, the "father" of American primatology. Lana learned to associate many lexigrams with their referents, including sequencing them and sometimes creating novel series. After Lana's success, two male chimpanzees, Austin and Sherman, began learning Yerkish with a very different approach. They were given greater flexibility to use lexigrams more functionally, and eventually were able to communicate with each other using lexigrams, and even learned to request tools from each other using symbols so they could use the tools to obtain tasty rewards. Another offshoot of the LANA Project, under Dr E.S. Savage-Rumbaugh, used the other species of chimpanzee known as the bonobo. This project's most famous subject is Kanzi who learned to use lexigram symbols by watching other bonobos and humans use the keyboard. Kanzi has also shown the capacity to learn to understand spoken English, a capacity previously reported in many historical accounts of chimpanzees, and now confirmed in bonobos also.

KANZI COMMUNICATES

Kanzi learned to communicate with Dr Savage-Rumbaugh using lexigram symbols by watching other bonobos and humans using them.

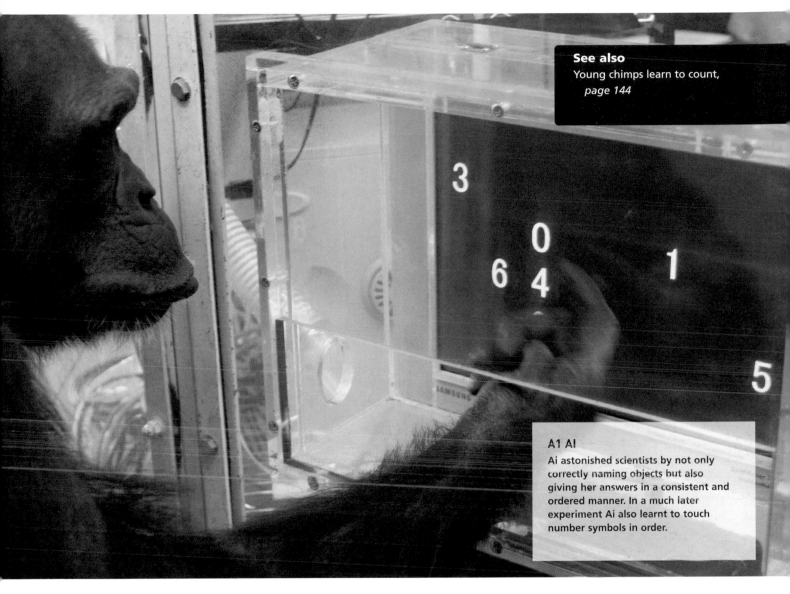

See also
Young chimps learn to count,
page 144

A1 Ai
Ai astonished scientists by not only correctly naming objects but also giving her answers in a consistent and ordered manner. In a much later experiment Ai also learnt to touch number symbols in order.

the correct colour label for all types of objects. And finally, Ai had to pass a test proving she could match the correct number symbol to a number of objects. After some time, Ai knew enough symbols to be challenged with the brand new test.

Three-in-one
Up to this point Ai had never been asked to tell her teachers the name of an object as well as its colour. Nor had she been asked to number and name the objects. Now, for the first time, Ai would be asked to label the object with the correct name, indicate its colour and tell her teachers how many objects were being shown, all at the same time. And because she had not previously been taught to give these three responses simultaneously, she had not learned how to order them. What she did surprised everyone.

The new test involved showing Ai several pencils (objects with which she was familiar), and they were either blue or red, in groups of 1–5. Ai got full marks. She used the correct object name, number and colour. Furthermore, she used her own order for each different set of objects. First, she gave the symbol name, then the colour and finally the number. Name, colour and number, and always in that order, even though she hadn't been taught to do that. Though the reasons remain a mystery, that's how Ai always organized her answers.

AN ORANG-UTAN LEARNS TO SIGN

NATURAL HABITAT
Borneo and Sumatra

APE LANGUAGE PROJECTS TRADITIONALLY USED FEMALE CHIMPANZEES UNTIL DR LYN MILES BEGAN PROJECT CHANTEK WITH A YOUNG ORANG-UTAN, AND SHE TAUGHT HIM TO COMMUNICATE WITH SIGN LANGUAGE.

Chantek (the word comes from the Malay language, and means "lovely" or "beautiful") was born in 1977 at a primate centre in Georgia. Miles received him when he was 9 months old and, from 1978 to 1986, he was totally immersed in a "signing environment", having human companions and carers with him at all times.

Chantek lived in a trailer on the grounds of the University of Tennessee at Chattanooga, and had daily signing sessions with Miles and other teachers and students who helped with his daily care. In most respects Chantek was raised very much like a human child, even receiving toilet training, and was assigned chores around the trailer just like the teachers and students. He earned an allowance of metal washers that he liked to spend at a fast-food restaurant and, furthermore, liked to clean up the area where he slept, sometimes lending a hand with the cooking. When he was still young, Miles took Chantek on outings to playgrounds and nearby lakes, and to a mountain where he could watch the hang-gliders sweeping through the sky.

Signing and understanding

Chantek learned several hundred signs, and also learned to understand spoken English as well as American Sign Language (ASL). But it's one thing to use ASL to name or label things, and another to understand or comprehend the use of signs by other individuals. Yet Chantek eventually became very proficient in making signs and understanding them during his sessions with his teachers. He developed these skills within the same developmental timeframe as children, and even invented

EXTINCTION

Orang-utans are now found only on two islands in Indonesia: Borneo and Sumatra. (The name "orang-utan" actually comes from the Malay and Indonesian language; "*orang hutan*" means "man of the forest".) These orang-utans are a subspecies of ape that have developed separate physical and behavioural characteristics. Both have arms that are twice as long as their legs, a fitting adaptation for their arboreal (tree-living) habitat in the wild. But because their natural environment is a rainforest, both kinds of orang-utans are highly endangered. Forest logging in Borneo and Sumatra has left extensive areas completely denuded of trees and other vegetation, forcing the orang-utans to move to other areas with a supply of fruit and foliage, which they desperately need to survive. Efforts are underway in several areas on both islands to preserve as much forest as possible for the few remaining wild orang-utans, or the two subspecies could face extinction in the next 50 years.

SIGN LANGUAGE INNOVATOR
Chantek the orang-utan spent the first nine years of his life immersed in a signing environment. He became so proficient at sign language that he invented some signs of his own.

See also
The great ape debate,
page 100

some signs of his own when he encountered new objects. For example, he apparently referred to contact lens solution as "eye drink", and also used descriptive adjectives with some names, for example "red bird".

Chantek's teachers provided him with a great number of opportunities. He did other types of play, just like children of about the same age. He loved to be tickled and chased, and enjoyed painting and other types of crafts. But, as Chantek got older, he physically outgrew his home. Eventually, he was relocated to the primate centre in Georgia (where he was born)

for a period of time, before being happily transferred to a special habitat that is part of Zoo Atlanta. There, Chantek roams within a naturalistic environment, with grass and trees to climb, and more freedom. He continues to sign regularly with zoo staff and Dr Miles is able to visit him.

ROCKY, THE SEA LION WITH A LOGICAL APPROACH

A REMARKABLE PROJECT WAS INITIATED IN 1978 WITH THE UNPRECEDENTED AIM OF TEACHING GESTURED AND VISUAL SYMBOLS TO CALIFORNIA SEA LIONS.

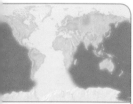

NATURAL HABITAT
Sub-arctic to tropical waters of the global ocean with the exception of the Atlantic Ocean

During the past three decades, there have been several successful research projects investigating the capacity of three different ape species, dolphins, sea lions and an African grey parrot to acquire an understanding of abstract symbols. And that simply means understanding a symbolic representation for objects, different foods, names of people in their environment, colours and a range of concepts involving similarity and difference. Among these innovative projects is the work of Dr Ronald Schusterman and his first pupils, Rocky and Rio, both California sea lions.

Schusterman began his scientific career in 1960 studying gibbons, monkeys and chimpanzees at the Yerkes Laboratory, Orange Park, Florida, before moving to Stanford Research Institute. He is credited with founding the North American laboratory that specifically focused on the behaviour and physiology of sea lions and seals. Ultimately, Schusterman moved his research program to the Long Marine Laboratory of the University of California at Santa Cruz. Several years later, inspired by the work with gestures and dolphins that Schusterman had seen at the laboratory of Dr Lou Herman in Hawaii, he decided to see what he could teach his sea lions using a similar approach.

Smart responses

Rocky was 10 and Rio 3 years old when they began their training with gestured signals. The early gestures that they were taught referred to objects that they could retrieve from their pool, such as a frisbee, ball and large cone. Action gestures were added to their repertoire, and the sea lions were soon responding to a sequence of signals that required them to perform an action on, or with, an object. For example, Rocky could respond to her trainer's gestured command "white ball–flipper–touch" and swim to the white ball, touching it with her tail. Rocky and Rio had to pay attention to all three gestures to choose the correct object, and then complete the action. Rocky also learned other terms so that she could take one object to another one in the pool, which suggests that she could comprehend simple sentences.

See also
Ask a dolphin and you'll get the right answer, *page 156*

QUICK LEARNERS
Along with apes, dolphins and parrots, sea lions have long been the subject of research into animal intelligence.

"WIN/STAY, LOSE/SHIFT"

This important concept is a crucial part of many animal-teaching programs. What it boils down to is this: if the animal answers correctly the first time when given two possible choices, giving it a 50/50 chance of being right, it should stay with that same choice the next time the same choice occurs. If, however, it is wrong when faced with these the two options, it should shift, or switch to the other alternative when the situation next occurs.

EQUIVALENCE

Scientists wanted to find out if sea lions were capable of understanding and remembering arbitrarily assigned pairs of symbols, and then using this information to make logical connections when presented with new symbols. Rocky learned to pair random images such as a black horse and a black teapot.

1| Rocky is shown a familiar central symbol (a black teapot) which was previously arbitrary paired with a black horse symbol and is now flanked by two new symbols.

2| She is then invited to choose one of the symbols from either side of the central symbol to win a treat.

3| Having failed to win the treat with her first choice, Rocky chooses the correct symbol on her second attempt.

4| She is rewarded with a fish.

5| In the next stage of the test, the central symbol is replaced with its original paired symbol.

6| Rocky has to make the link between the original pair and the new image that wins her the treat.

Schusterman and his students wondered whether they could discover if Rocky was actually "thinking" like a human, and came up with a task to challenge the sea lions. They wanted to see if Rocky could understand certain types of logical relationships that they thought was possible only through language. Schusterman believed that Rocky's ability to understand gestured signs, and the objects and actions they represented, was supported by a basic type of learning mechanism that was the root of more complex information processing.

Logical thought

So the teachers devised a set of graphic images, including Arabic numerals. Then a wooden apparatus composed of three large panels was built, each with a square window in the middle that could be opened to display a symbol. In front of the panels was a chin rest for Rocky to ensure she was looking at the middle panel at the start of each trial. Rocky was then taught to associate two abstract symbols that the experimenters had arbitrarily paired together. When the experiment began, the centre window was opened to reveal one symbol, followed by the two side windows, and Rocky had to look at the latter two, and select one by touching it with her nose. Early in her training, Rocky had to choose simply by chance, and was rewarded with a tasty fish if she chose correctly. Eventually, Rocky learned the process of trial-and-error, understanding that if she was wrong with her first choice, the next time that she saw the same pair again, she should pick the other option. If by chance she chose correctly the first time, Rocky had to remember that she had been rewarded for that response, and choose it again the next time. This strategy is described as "win/stay, lose/shift" in animal learning.

Over time, Rocky learned the associated pairs of symbols, and was soon ready for the real test when she'd be presented with symbols that had never been paired together before. Would Rocky make the logical connection between the new pairings? For example, suppose Rocky had learned to pair the black silhouette of a horse with the black image of a teapot, and also learned that when the teapot was shown she should choose the scissors symbol. Now, when the horse symbol was presented, she'd have only two choices – the scissors and a key. Rocky was meant to choose the item that had previously had a logical relationship with the horse and teapot, the scissors, because of its shared association with the teapot. Easy!

Rocky did not hesitate in responding correctly to all the new combinations. She immediately grasped the logical relationship among the symbols. In effect, she was linking symbols together, even when she hadn't been taught that they were pairs, demonstrating a logical process called "equivalence", previously thought to be unique to humans, and requiring language to function. Consequently, Rocky demonstrated that she was able to "think" in ways rather like a human.

CHAPTER SEVEN
COOPERATION AND ALTRUISM

Perhaps the most significant difference between humans and other animals is our ability to empathize with our own species. Indeed, current thinking among biological anthropologists is that the demand for cooperation in a complex social structure was a major impetus for the development of higher intelligence in humans. However, strong evidence for empathetic behaviours such as food sharing and cooperative hunting have been documented among other species such as bats, capuchins, chimpanzees, baboons, wild dogs and lions. In some cases, more highly sophisticated exchanges among group members reflect reconciliation and consolation, reflecting some types of empathetic responses previously thought only to be expressed by humans.

"I'LL SCRATCH YOUR BACK NOW, IF YOU'LL SCRATCH MY BACK LATER"

DARWIN'S IDEA THAT NATURE PROMOTES THE "SURVIVAL OF THE FITTEST" SUGGESTS THAT ALL ANIMALS HAVE EVOLVED TO BE SELFISH AND COMPETITIVE. HOWEVER, THERE IS EVIDENCE OF ANIMALS USING INTELLIGENT SOCIAL TACTICS THAT PROMOTE HARMONIOUS SOCIAL RELATIONSHIPS.

NATURAL HABITAT
Central and
South America

When Darwin suggested that the natural world has evolved according to the principle of the "survival of the fittest", he assumed this meant that animals would be driven by entirely selfish motives. Each individual would be competing with its fellows for a limited set of resources. According to Darwin, such a state of affairs would produce nasty, aggressive, dog-eat-dog societies.

Survival of the relatives

While there is a great deal of aggression and competition in the animal kingdom, there are also many examples of friendly, cooperative and even apparently self-sacrificing behaviour. For example, many animal species give alarm calls when they detect a predator, which immediately places the caller in danger by attracting the predator's attention. Yet many species of birds, rodents and primates do just this.

Surprisingly, researchers have shown that there might still be biologically selfish motives for such alarm calling. For example, Belding's ground squirrels are much more likely to call when their audience consists of close relatives rather than distantly related individuals (see page 66). The evolved tendency to promote the survival of one's relatives is called "kin selection". We share more copies of our genes with our close relatives than with strangers. By helping our relatives, we are therefore promoting our own long-term genetic survival. Hence, a great deal of alarm calling might be genetically hardwired, rather than involving either completely selfless motives or intelligent decision-making.

More blood?

Yet, animals do not always just help close family members. There are some animal species that appear to make personal sacrifices for non-kin. Perhaps the most extraordinary example comes from vampire bats.

RECIPROCAL ALTRUISM

One classic observation of reciprocal altruism was provided by Craig Packer's research on the consort behaviour of wild olive baboons. When female olive baboons become sexually receptive, their bottoms swell to quite grotesque proportions. When this happens, the adult male baboons begin to compete with one another in an attempt to gain exclusive sexual access to the female. They try to form what is called a "consortship" in which a single male herds a fertile female away from the rest of the males. The other males will often follow a consorting couple and try to steal the female away. Packer noticed that males were more likely to steal a female if they worked together in pairs. Two males would follow a consorting male, continually harassing him. If the consorting male lost his cool and suddenly attacked one of his harassers, the third male would often use this opportunity to slip in and steal the female. Packer was able to show that the male baboons who worked together seemed to be taking turns. If on one occasion baboon A gained the female, then on a subsequent occasion baboon B would gain her. In other words, they seemed to be reciprocating. But despite Packer's exciting preliminary data, no one has been able to replicate these findings.

BLOOD TIES
Vampire bats, which need to feed on blood regularly to stay alive, often share food with other bats in the colony that have not been able to feed.

One researcher suffered gruelling conditions in order to closely study wild vampire bats. The bats lived in colonies that roosted in the hollows at the base of massive trees. To collect his data, the researcher, Gerald Wilkinson had to crawl into the base of these hollows and lie on his back, suffering not only the awful stench of the bats' guano pile but also the unwanted attention of the hungry little vampires. But his suffering was not in vain. He discovered an amazing pattern of blood-sharing among the bat colony.

Vampire bats must feed on fresh blood at least once every three days, otherwise they starve. Young bats often fail to feed and, when this happens, they desperately beg other members of their colony for supplies. Bats who have successfully fed will often regurgitate blood to their begging companions. Wilkinson suspected that although much of the blood sharing was among close family members, it might also take place between non-kin. He devised a clever way to test his theory. He caught and formed colonies of distantly related wild bats in his laboratory. He then removed some bats and prevented them from feeding. When returned to their captive colony, other bats would often share with them even though they were not close relatives. Furthermore, the bats seemed to keep track of who had shared with them in the past, and they were much more likely to reciprocate with those who had been generous to them on a previous occasion.

The pattern of sharing exhibited by vampire bats adheres to Robert Trivers' concept of reciprocal altruism. Trivers (a wide-ranging anthropologist and sociobiologist) suggested that the tendency to show self-sacrificing behaviour to non-relatives might evolve if individuals reciprocate favours with one another. It is a case of, "I'll scratch your back now, if you scratch my back later". However, the tendency to perform reciprocal altruism is likely to be rare since it requires quite complex intellectual abilities. The reciprocators must live in stable social groups, be able to recognize individuals and remember and act upon the details of previous social exchanges. There is, in fact, very little strong scientific evidence of reciprocal altruism in wild animals.

Despite the fact that reciprocal altruism has proved difficult to find in animals, scientists have identified many behaviours in several different species that seem to function to promote harmonious social relations. And some of these behaviours seem to require some degree of intelligence.

See also
Ground squirrels look out for their own, *page 66*

CHIMPS "KISS AND MAKE UP" AFTER A FIGHT

THE MEMBERS OF A SOCIAL GROUP COMPETE FOR THE SAME RESOURCES, WHICH INEVITABLY LEADS TO CONFLICT. HOWEVER, THERE ARE CERTAIN COMPLEX SOCIAL TACTICS USED BY SOME SPECIES THAT SEEM TO HELP EASE TENSION AND PROMOTE HARMONIOUS SOCIAL RELATIONSHIPS.

NATURAL HABITAT
Western and Central Africa

Living in a social group is bound to produce a certain degree of conflict. When a group consists of members of the same species, they inevitably compete over the same resources: after all, they eat the same kinds of food, need the same kind of shelter and are attracted to the same mates. Despite this inevitable competition, there are benefits to living in a social group, such as greater protection from predators, greater chances of finding and securing food and increased access to fertile members of the opposite sex. Group members would waste a great deal of energy if they were constantly in open competition with one another. Hence, there is a balance to be struck between individuals securing as much of the limited resources as possible, while maintaining stable social relationships.

Dominance hierarchies

One way in which many social species seem to strike a balance between competition over resources and maintaining reasonably harmonious relationships is to form dominance hierarchies. In these hierarchies, each member of the group knows its place and will defer to higher-ranking individuals. The behaviours that maintain these rank structures are probably rather hardwired and do not need a great deal of intelligence to support them.

However, there are certain types of behaviours that have been observed in a few species of chimp that would appear to function to promote social harmony, and they may involve a degree of intellectual sophistication. One such behaviour has been dubbed "reconciliation".

When Frans de Waal was a postgraduate student in Holland, he studied a large group of captive chimpanzees at the Arnhem Zoo. The chimpanzees lived on an island surrounded by a moat of water. De Waal noticed that the Arnhem apes would often approach each other in a friendly manner shortly after a fight. One of the former combatants would extend his or her hand towards the other. They would then embrace or even kiss. De Waal wondered whether these patterns of behaviour were the chimpanzees' way of reconciling their differences. To continue fighting when the chimps lived in such constant close contact with each other would have been potentially very destructive.

Making up after a fight could serve to ease the tension quickly and, thereby, prevent a fight from escalating out of control.

When de Waal first presented his ideas on chimpanzee reconciliation to his academic colleagues, they were very sceptical. They felt that it was highly unlikely that any animal species, even our closest genetic relative, the chimpanzee, would possess such complex social abilities. In order to convince them, de Waal devised a clever method to validate his ideas. He was able to observe the Arnhem Zoo chimps from tall observation towers wherever they went on their island home. Every time there was a fight, he watched the chimps that were involved for the next 10 minutes to see if they approached each other in a friendly manner. He then collected another set of data to

See also
Wild chimpanzees: masters of communication, *page 76*

compare with these post-fight periods. He watched individual chimps for 10-minute periods when the apes had not been involved in any kind of a fight. He was therefore able to show that the chimps were much more likely to approach, embrace, kiss, pat and groom each other within the first 10 minutes after a fight than at any other period when they had not been fighting. Hence, de Waal provided strong statistical evidence to support his claim that the chimps were especially friendly to each other after a conflict.

De Waal's concept of reconciliation, and his scientifically rigorous method for demonstrating it, caused a great deal of excitement and interest among other animal researchers. Researchers working with many different species have used these methods to test whether their subjects also reconcile with one another. To date, reconciliation has been observed in many species of primates, and even in some non-primate species including feral sheep, spotted hyenas, lions, dolphins, dwarf mongooses and domestic goats.

Consolation

Nevertheless, there are still some forms of behaviour that chimps exhibit after a fight that seem extremely rare, if not absent, in other species. According to one report, there's a pattern of behaviour in chimps that has been labelled "consolation". Occasionally, after a fight, a third party who had not been directly involved would approach, hug, kiss or groom the winner or loser. The report suggests that consolation is similar to reconciliation because it functions to ease group tension. However, consolation goes one step further than reconciliation in its social complexity, since the consoler was not directly involved in the fight but still goes out of his or her way to try and calm fellow group members.

APE WATCHING
Scientists record chimpanzee behaviour at Burgers Zoo in Arnhem in the Netherlands.

ARBITRATION

De Waal describes an even rarer pattern of behaviour in chimpanzees, which he calls arbitration. This involves a third party, who was not directly involved in the original fight, reconciling the two former combatants. One of the chimps in the Arnhem group, an old female called Mama, was a particularly skilled arbitrator. She seemed very keen to promote harmonious social relations between the large, dominant adult males in her group. If two of these males had been fighting, Mama would sometimes approach and groom either the winner or loser. Grooming is a relaxing activity that often serves to calm individuals down. On a few occasions, de Waal observed that after Mama had groomed one of the males for a while, she would gently grasp his hand or arm and lead him over to his former opponent. She would then groom the other chimp while the first male groomed her. Eventually, Mama would extricate herself from the grooming huddle, leaving the two males grooming each other. Such social skills seem extraordinary, and it has lead researchers to suggest that perhaps chimpanzees, if not other species, possess a capacity for empathy.

MAKING UP

Chimpanzees reconcile by grooming, patting and even kissing after a fight.

1| A fight breaks out between two members of the group (A and B).

2| A third member of the group (C) approaches and grooms one of the fighters (A).

3| The peace-maker (C) brings the fighters together, still grooming the first (A), while inviting the second fighter to join in the grooming (B).

4| After a while, the peacemaker (C) slips away discreetly, allowing the two enemies (A and B) to make up with one another.

NATURAL HABITAT

Dogs and mice:
Worldwide

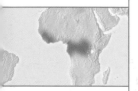

Chimps: Western and
Central Africa

Dolphins: Worldwide,
mostly in the shallower
seas of the continental
shelves

YAWN AND THE WORLD YAWNS WITH YOU – EMPATHETIC RESPONSES

IT HAS LONG BEEN HELD THAT EMPATHY IS A UNIQUELY HUMAN CAPACITY. HOWEVER, THERE IS INTRIGUING EVIDENCE FROM OTHER ANIMALS THAT THEY TOO MAY HAVE THE ABILITY TO UNDERSTAND AND RESPOND TO THE DISTRESS OF OTHERS.

Empathy is the ability to share and respond appropriately to the feelings of others. At its simplest, an individual might be affected by the distress of others by also becoming distressed. Such a response is called emotional contagion. It is a fairly automatic process that does not require a great deal of intelligence. Even newborn human babies will start crying in response to hearing other babies cry. But, at its most cognitively complex, empathy involves taking the mental perspective of another individual and responding to it in a way that specifically meets the other's needs. So a child might respond to the distress of another child by summoning the other child's mother, or by retrieving and offering the child a favourite toy. The comfort that is being offered is directly tailored to the other child's needs.

COMPLEX EMPATHETIC RESPONSES

Two studies on monkeys suggest that they may be capable of more complex responses. In one bizarre experiment monkeys had a horrible choice: pull a chain and receive food as a reward but in so doing give another monkey in an adjacent cage an electric shock. The monkeys showed greater empathy than the humans who devised these experiments. They very quickly began to refuse to pull the chain once they learned it caused another monkey pain. Their behaviour seemed to go beyond emotional contagion, since they showed a willingness to sacrifice their own wants in order to avoid causing suffering to others.

DOG TIRED
Dogs have been observed to yawn when humans around them do.

I FEEL YOUR PAIN
Mice respond to the pain of other mice, especially if they have been cage-mates for some time.

In a longitudinal study of human children, it was found that infants begin to show appropriate empathetic "comfort offering" from 12 to 24 months old. When the infants' mothers pretended to cry, many of them tried to offer comfort by hugging, patting and offering toys. But while humans show a capacity for empathy at an early age, there is little evidence of empathy in other species.

Anecdotal evidence
Most of the evidence for empathy in non-human animals is anecdotal. Perhaps some of the most famous examples come from dolphins, which are renowned for responding to distress in human swimmers. Stories abound of dolphins saving humans from drowning. However, it is almost impossible to confirm the veracity of such stories. Maddalena Bearzi has been studying wild dolphins for over 10 years. She was very sceptical about such stories until she witnessed an incident for herself. She was following a group of dolphins that were hunting in shallow waters. Suddenly the dolphins turned and began to swim very quickly into deep water. When Bearzi followed in her speedboat, she found that the dolphins had surrounded a young woman who was drowning in a suicide attempt. After Bearzi had pulled the young woman aboard, she realized that the dolphins had gone. Bearzi was uncertain how to interpret the dolphins' behaviour. Were they showing empathy for a struggling swimmer, or were they merely curious about this strange, flailing being?

Full-blown empathy
The problem with anecdotes is although they are often fascinating, it can be difficult to interpret them. Frans de Waal recounts many anecdotes of apparently empathetic behaviour in chimpanzees. For example, he describes the behaviour of a female bonobo, Zuni, who lived in the San Francisco Zoo. One day a bird flew into Zuni's enclosure and injured itself by flying into a large glass window. Zuni very gently picked the bird up, stroked it and carried it to the top of one of the climbing frames. There she turned the bird so that it faced away from her, held it up by its wings and launched it into space. Unfortunately, the bird was too stunned to fly and fluttered back down to earth. Zuni retrieved it and gently carried it for the rest of day until the bird had recovered sufficiently to fly away.

De Waal considers Zuni's behaviour as an example of full-blown empathy: she responded in a way that was relevant to the perspective of the bird. She carried it to a high vantage point

to enable it to fly, and held it up facing away from her before releasing it. Yet Zuni could also have been responding to the bird as if it were a delicate toy. She may have had no real concern for its welfare, or understanding of its predicament. It is impossible to tell from one isolated instance whether bonobos possess the capacity to show empathy for such a different creature.

Mice and dogs
Studies on mice have found that they respond to the pain of other mice, especially if the mice have been cage-mates for a period of time. The mice writhed and licked their paws on seeing a former cage-mate in pain. They also seemed more sensitive to their own pain when in the presence of another suffering mouse, and that applied even if the two mice were experiencing different kinds of pain, for example from a heated floor and artificially induced nausea. The reactions of the mice were consistent with emotional contagion.

Recently a team of scientists from Birkbeck College, University of London, reported a different type of contagion in domestic dogs. They found that when a human yawned in an exaggerated manner in front of pet dogs, it significantly increased the likelihood that the dogs would also yawn. The researchers discussed the dog's behaviour in the context of empathy. Contagious yawning involves synchronizing one's internal state to that of another. Being rather like emotional contagion it might be a simple precursor to more complex empathetic responding.

See also
Chimps use mirrors like we do – to see how they look, *page 108*

NATURAL HABITAT

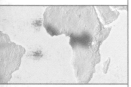

Chimps: Western and Central Africa

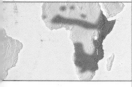

African wild hunting dogs: Eastern and Southern Africa

Capuchins: Central and South America

A DOG'S DINNER IS A SHARED AFFAIR

VERY FEW SPECIES SHARE FOOD WITH NON-RELATIVES, WITH FOOD SHARING BEING PARTICULARLY RARE IN NON-HUMAN PRIMATES. HOWEVER, BOTH CAPUCHIN MONKEYS AND CHIMPANZEES HAVE BEEN OBSERVED ACTIVELY SHARING FOOD WITH OTHERS.

Humans share food with one another on an almost daily basis. It comes so naturally to us that it never strikes us as odd. However, food sharing is a very rare pattern of behaviour in primates, and very few voluntarily hand food to another individual, even their own offspring. In contrast, food sharing in humans seems to be an ancient pattern of behaviour. Hunter–gatherer peoples, who live entirely by harvesting wild plant foods and wild game, follow a way of life that paleontologists believe was typical of our Stone Age ancestors. All the hunter–gatherer groups ever studied exhibit very strong traditions of sharing their food with the rest of the group. Not only do they share with close family members, but also with relative strangers.

Lots of other species share food with their young, but sharing with non-family members is much less common. African wild hunting dogs carry partly digested meat inside their bodies – often for long distances – from a kill back to their den and regurgitate it for other pack members who have usually stayed behind on puppy-sitting duty. The dogs will share with other pack members even when they are non-kin.

Among primates, two species stand out for their tendency to share food: capuchin monkeys and chimpanzees. Most of the sharing one finds in capuchins is probably best called "tolerated theft". The capuchins are very tolerant of each other. They will allow other monkeys to approach them when they are feeding, and will even sometimes tolerate another monkey picking up scraps of food or occasionally taking pieces of food right out of their hands.

Sharing in captivity

Frans de Waal has conducted studies on food sharing in capuchin monkeys. He attached testing chambers to the front of the monkeys' home enclosure, and would call the monkeys by name, they would then voluntarily climb into the chambers. De Waal placed two chambers adjacent to each other so that the two monkeys were divided only by a single wall of wire mesh. He then fed one of the monkeys but not the other with fruit, and discovered that they began to share with one another. The monkeys that received food not only sat and fed immediately next to the mesh divider so that it was easy for a neighbour to reach through and pick up scraps, but they would also push pieces of food through the mesh wall into the adjacent chamber.

DOGGY BAG
African wild hunting dogs swallow meat and carry it inside their bodies back to the den. They then regurgitate a portion of partially digested meat for the lead female's puppies and their babysitter.

See also
Yawn and the world yawns
with you – empathetic
responses, *page 176*

There was also some evidence of reciprocal altruism. The monkeys were much more likely to push food through to a neighbour if that individual had shared food with them on a previous occasion.

Although such sharing is easy to induce among captive capuchins, there is little evidence of sharing in the wild. Partly this is because plant foods tend to be evenly distributed, and there is no need to hand pieces to other monkeys who can just as easily help themselves to what is available. In fact, human hunter–gatherers tend not to share plant foods with one another either. Hunter–gatherer women collect vegetables and berries for the family pot rather than for everyone in the group. The main concentration of food sharing among hunter–gatherers involves meat. Hunting in all hunter–gatherer peoples is the domain of men, and when an animal carcass is brought back to camp it is shared with everyone in the group.

Meat-sharing chimps

The only other primates that have been seen to share in the wild are chimps and, just like human hunter–gatherers, this usually involves meat. Chimps are one of the few primate species that hunt and eat other animals. If the hunt is successful, there is a great deal of excitement surrounding the kill. Other chimps will vehemently beg individuals with chunks of carcass for a piece. Under these circumstances it is not unusual to see a chimp break off a small piece of meat and hand it to a begging companion.

Chimpanzees seem to be rather tactical in the way they share meat. The male chimps do most of the hunting and once the prey, which might be a colobus monkey, a small antelope or a bush pig, has been caught and killed, they tend to share mostly with other males who also took part in the hunt. If a high-ranking male turns up after the hunt, he will be less likely to gain a share of the meat than lower ranking fellow hunters.

Not only will the hunters share with one another, but often with adult females. It has been suggested that males might have an ulterior motive for their generosity to females. If a male shares meat with a female, the next time she becomes sexually receptive she might be more likely to mate with him. Although the meat-for-sex hypothesis seems reasonable, when systematic observational data were collected, no relationship could be found between meat sharing and future mating success. It is therefore somewhat of a mystery why males share meat with certain females, but not others.

WORKING TOGETHER – IS IT TEAMWORK OR JUST A BUNCH OF ANIMALS?

NATURAL HABITAT

Lions: Sub-Saharan Africa

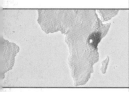

Baboons: East Africa

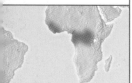

Chimps: Western and Central Africa

Capuchin monkeys: Central and South America

MOST ANIMAL COOPERATIVE BEHAVIOUR, SUCH AS GROUP HUNTING IN LIONS, APPEARS TO BE OPPORTUNISTIC RATHER THAN INTENTIONALLY COORDINATED. HOWEVER, THERE IS SOME EVIDENCE FROM PRIMATES THAT THEY DO MONITOR EACH OTHER'S BEHAVIOUR WHEN JOINTLY WORKING ON A PROBLEM.

Cooperation means two or more individuals working together towards the same goal, thereby gaining an overall net benefit exceeding what they would have achieved had they worked alone. Perhaps one of the most famous examples of cooperation in animals is group hunting in lions. Lions are unique among the big cats because they live in groups. Research has shown that when lions hunt in groups, instead of individually, they bring down larger prey with a higher success rate. But while they cooperatively hunt, this does not mean that their hunting tactics are very intelligent. One observer noted that the way they space themselves as they approach the prey seems to be more opportunistic than intentionally coordinated.

The Pumphouse Gang

The hunting tactics used by baboons – one of the few primates that have been observed hunting and eating other animals – also seems to be mostly opportunistic. Shirley Strum studied the hunting behaviour of a group of baboons called the Pumphouse Gang, so named because they liked to hang out near a water pumping station. When she first observed the baboons hunting, one individual would start to hunt and when the prey (usually a hare or baby antelope) began to flee, other males would join in. But once a baboon caught the prey, he would never share it. If a female managed to get some meat, usually one of the large males would aggressively take it.

One day a new male, and keen hunter, named Rad, joined the gang. The number of hunts increased dramatically, and the baboons began to show signs of being somewhat more cooperative. Rad would assume a distinctive alert posture when he spotted likely prey, and the other baboons were quick to

FAMILY PRIDE

Lions live in large social groups. The closely related lionesses within a pride will hunt cooperatively, feasting on the prey together.

See also

Lions show roar talent,
page 132

THE ROPE TEST

Two capuchin monkeys work together to pull a weighted platform into reach. Two dishes of food are placed on the platform. The monkeys can't pull the platform and reach the food by working independently. If an opaque barrier is placed between them, then they can't sufficiently coordinate their pulling to reach the food. Therefore, they show coordinated behaviour which is an important aspect of cooperation.

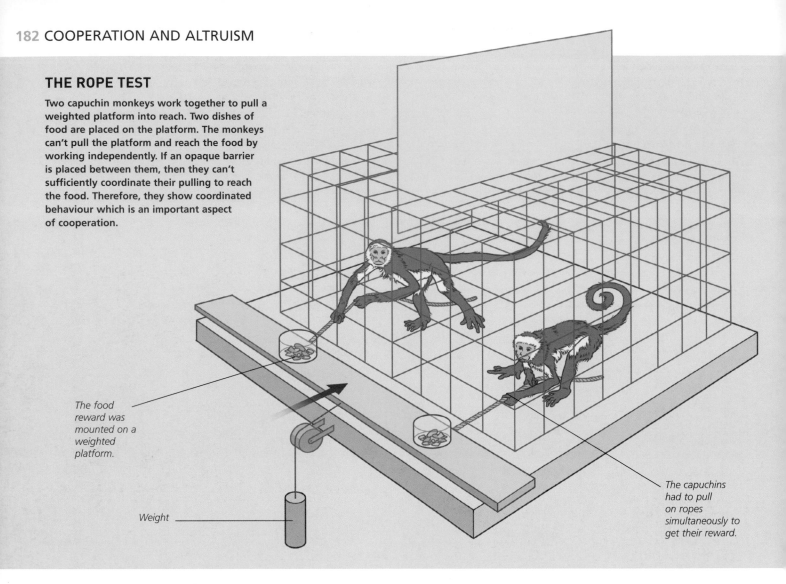

The food reward was mounted on a weighted platform.

Weight

The capuchins had to pull on ropes simultaneously to get their reward.

respond and approach. Soon, more individuals would join the hunts, including females. However, if a female was able to catch the prey, she was much more likely than before Rad's arrival to keep and eat the prey herself. Despite the increased coordination, the Pumphouse baboons' behaviour was still more opportunistic than cooperative.

Cooperative chimps

In contrast, other research suggests that the chimpanzees of the Tai forest intentionally cooperate when they hunt. They give a distinctive call to signal the start of the hunt, then the chimps assume different roles: drivers push the prey forwards, chasers try and grab it, blockers stop the prey from escaping and ambushers wait in thick foliage to pounce. The chimps also share the meat with one another according to who has taken part in the hunt; an individual's dominance ranking is not important. Other researchers refute these claims, suggesting that the hunt might be no more cooperative than that of lions. The problem is that it's

very difficult from observations in the wild alone to establish how intentionally cooperative the chimps are being. However, there have been a number of experiments on cooperation in non-human primates, including chimpanzees.

The first study involved placing a huge block of concrete outside the cage of two young chimps. Bananas were placed on top of the block, and ropes were then attached to it and threaded through the bars of the chimps' enclosure. The block was too heavy for one of the chimps to pull into reach on his own. To reach the food they had to work together pulling simultaneously. When the apes were able to do this it was claimed that they were intentionally cooperating. Yet, it is difficult to tell whether the chimps were intentionally coordinating their behaviour or whether their synchronization was more opportunistic. Certainly the film of the chimps' behaviour suggests that it was intentional. One of the chimps looks as if he's getting bored with the task and starts to walk away. His companion approaches him, puts his arm around his cage-mate's shoulders and coaxes him back to the ropes. They then pull the block into reach and the bored chimp walks away again, apparently content that his companion got the whole banana.

The second rope test

A more recent version of the rope-pulling task again tested for cooperation in capuchin monkeys. The monkeys were presented with food on a weighted platform outside their cage. To reach the food they had to pull on ropes simultaneously, which they did successfully. In a slightly smarter test to see if synchronization was based on chance, the monkeys were separated from each other by a wall of wire mesh. When the mesh was replaced by a solid, opaque barrier, the monkeys were unable to coordinate their behaviour. So it seems that the capuchins were monitoring each other while pulling, even if this was not immediately obvious to the human observers.

A later experiment replicated the rope-pulling experiment with chimps. This showed that the chimps were able to cooperate, but they were less able to share than the capuchins. If they had tense relationships before they started the tests, they found sharing the food prize, even if it was dispersed on the board, difficult to do.

So, there is evidence that some species of primates are capable of intentional cooperation. Furthermore, more tolerant species can promote cooperative exchanges by sharing their rewards.

CHIMPS AND CAPUCHINS

Raphael Chalmeau devised a special task to test for cooperation in chimps. He presented a group of six chimps with a pair of levers spaced far apart, which had to be pulled simultaneously by two different individuals for the apparatus to release food. The adult male and a juvenile female succeeded in working together. The female seemed to pull randomly on the levers, but the male carefully watched her and synchronized his pulling with hers. However, he was a bit of a bully and refused to share the food reward, at which point the female refused to help. Further research replicated the levers task with capuchin monkeys. The monkeys also succeeded, but they did not seem to monitor each other while pulling the levers, indicating that synchronized pulling seemed to occur by chance. Since the first study was rather marred by the bullying male, two researchers – de Waal and Berger – tried to discover if their capuchin monkeys would fare better. They showed that if, after pulling on the ropes together, one of the capuchin monkeys gained the food but the other did not, then the first monkey would be more likely to share with its companion.

WORKING TOGETHER
Chimpanzees have demonstrated cooperative sychronized behaviour in pursuit of a goal. However, it is not clear if this is intentional or opportunistic.

MONKEYS ARE QUICK TO SPOT AN UNFAIR DEAL

NATURAL HABITAT

Capuchins: Central and South America

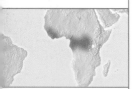

Chimps: Western and Central Africa

CAPUCHIN MONKEYS AND POSSIBLY CHIMPANZEES SEEM TO POSSESS A SENSE OF FAIRNESS. CHIMPANZEES CAN ALSO TELL THE DIFFERENCE BETWEEN ACCIDENTAL AND DELIBERATE ACTIONS, AND WILL SELFLESSLY RETRIEVE AND RETURN AN OBJECT ACCIDENTALLY DROPPED BY HUMANS.

Human beings will express a sense of righteous indignation if they perceive that they have been treated unfairly. So they may work quite happily until they learn that someone else is receiving higher wages for the same job. Although a sense of fairness is strong in humans, it might seem unlikely that any animal species would possess similar sensibilities. Nevertheless, tests have shown that capuchin monkeys and chimpanzees might also possess a sense of fairness.

A fair deal

Researchers placed capuchin monkeys in two separate, adjacent, wire mesh chambers. The monkeys had learned to trade stones for pieces of food. One monkey in one chamber was offered cucumber pieces in exchange for stones. Then a monkey in the second chamber was offered grapes (far tastier to capuchins than cucumber) as an exchange. When the first monkey noticed that the neighbour was receiving a better deal, he refused to exchange any more stones. And some of the monkeys seemed so upset by the inequity that they retreated to the back of the chamber and curled up in a tight ball.

Now it might just be that simply seeing grapes would be sufficient to make the first monkey dissatisfied with the cucumber, in which case fairness wasn't an issue. Even if no other monkeys were present and just a bunch of grapes was placed in view, the monkey might have stopped exchanging stones for the cucumber. However, tests showed that if the monkeys were just being given food, they were happy to receive cucumber even if a neighbour was receiving grapes. So it really did seem that the monkeys expected to receive the same exchange deal. To be given a boring cucumber instead of grapes was not fair.

See also
A dog's dinner is a shared affair, *page 178*

TOO KIND

A recent set of experiments conducted shows that chimpanzees will selflessly help others. The chimps watched either an unfamiliar human or a genetically unrelated chimpanzee struggle with a problem, for example not being able to reach an object. The watching chimps spontaneously helped even in the absence of a reward, or when helping was costly in terms of exerting a reasonable degree of physical effort. The chimps seemed to understand the goal of the struggling chimpanzees and humans, and helped them to achieve it.

Chimp fairness

If capuchin monkeys seem to possess a sense of fairness, perhaps chimpanzees also do. Researchers carried out a similar test to the grape–cucumber exchange, and the apes behaved like the monkeys, except that the apes were more tolerant of a bad deal if they already had warm, friendly relations with their neighbour. It's worth stressing that capuchins live in a more equal and tolerant society than chimps, who live in groups consisting of shifting alliances and complex political relationships. So it shouldn't be surprising that the chimps' long-term relationships affect how they respond to an unequal deal during a short-term experiment. That was the theory, anyhow. A similar experiment gave contradictory results. Regardless of their relationship with a neighbour, this time the chimps just begged more strongly when the lucky neighbour received a better deal, instead of refusing to exchange. So it is difficult to tell whether chimps really do possess and act upon a sense of fairness.

Guilty or not guilty?

There are also contradictory results when trying to determine whether chimpanzees can distinguish between a deliberate and an accidental act. In one experiment, humans handed cups of orange juice to chimpanzees. On every occasion the juice was spilled on the floor. However, sometimes the researcher deliberately poured the juice on the floor, and other times "accidentally" dropped the cup. The chimps were then given the opportunity to choose who would hand them the juice. If they were taking accidental clumsiness and deliberate meanness into account, one might expect the chimps would prefer the person who accidentally dropped the juice. In fact they made no such choice.

However, another experiment did show that chimpanzees can discriminate between deliberate and accidental actions. Food was hidden in one of two boxes, and a researcher placed a small wooden cube on one of the boxes to indicate where the food was located. The chimps learned to point at the box with the cube on top, and when they did so were given the food inside. However, in some trials the cube was "accidentally" dropped onto an empty box and then moved to the correct one. And on other occasions, the cube was deliberately put in the right place, and was then "accidentally" moved to an empty box. The chimps were able to tell the difference between accidental and deliberate placing, and they pointed at the right boxes accordingly (2 and 3 year old children and orang-utans behaved in a similar way to the chimps on the same test). Interestingly, while the experiment showed that the chimps were able to tell the difference between the two actions, according to the juice experiment they do not necessarily use this information to make judgements about the character of a person.

JUSTICE FOR ALL
According to research scientists, capuchin monkeys possess a strong sense of equality.

GLOSSARY

Alarm barks, calls Vocalizations produced in response to the sight of a predator or other perceived danger.

Altruistic alarm calling Vocalizations given when danger is perceived, despite increased risk to the caller.

American Sign Language, Ameslan, ASL Standardized gestural language system used in the United States by the hearing-impaired.

Attribution of mental states The understanding that other individuals have "minds", and may possess different needs, beliefs and desires than one's own.

Bush meat trade Current poaching of African wildlife, including endangered gorillas, chimpanzees, and bonobos, for commercial selling of meat to Kenyan restaurants.

Comparative cognition Study of the cognitive abilities and capacities of non-human animals.

Comparative psychology Study of the learning and behaviour of non-human animals.

Conspecific Term that describes a member of the same species (eg, wolves hunt cooperatively with their conspecifics – other wolves).

Darwin, Charles Highly influential theorist and naturalist who provided the first comprehensive framework for the process of evolution, published as *The Origin of Species*, in 1859.

Differential altruistic calling Vocalizations given during conditions of potential danger, typically in the presence of kin, though not when non-relatives are present.

Dominance hierarchy Animal social structure that reflects relationships among group members based on rank.

Double-blind testing Experimental method during which one experimenter monitors one facet of an experiment, and a second experimenter manages the subject's responses, ensuring that neither experimenter influences the other's participation.

Emulation Behavioural response to observed actions of another individual that results in the observer's attention being drawn to other's behaviour, such that observer then reproduces the observed behaviour or actions.

Enculturation Enculturated rearing of a non-human animal, usually apes, under enriched environmental and social conditions similar to raising children.

Equivalence Logic relationship defining symmetry or similarity.

Extractive foraging Food-seeking behaviour that involves tool use for extracting food from an embedded source (eg, cracking open nuts using stones).

Fission–fusion society Dynamic animal social structure whereby community members do not stay in a static group, but leave for periods of time before returning.

Hominid General term for modern or extinct primates who walked upright (bipedal), including all species of the genus Homo (eg, Homo sapiens) and the Australopithecines.

Imitation Reproduction of observed behaviour(s) with absolute fidelity, at the first opportunity to perform the behaviours after observation.

Incest taboo Behavioural predispositions among animals that appear to be genetically programmed to ensure that inbreeding between parents and offspring does not occur.

Infrasound Extremely low-frequency sound that is below the level of human hearing which serves as communication mode for several large species, including elephants and whales.

Insight learning Type of learning once thought to differ from other types, characterized by the presumed sudden and spontaneous solution to the problem.

Mark-directed behaviour Behavioural responses to visible marks made by dye or other materials and placed on the skin of subjects; typically entails touching marks and/or smelling fingers after physical exploration of marks.

Mark Test Experimental task pioneered by Gallup to test for mirror self-recognition in animals by marking the face and other body parts with dye, then exposing the subject to a mirror.

Matriarch Female leader of a social group, often the eldest and most experienced.

Mirror-contingent behaviour, mirror-guided behaviour Behaviours that use mirrored feedback to orient touching or exploring parts of the body that are visually inaccessible without a mirror.

Mirror self-recognition (MSR) The ability to understand that a mirrored reflection is one's own face and body.

Musth Period of heightened arousal and sexual excitability related to mating in male elephants, resulting from a hormonal surge.

One-to-one correspondence Learned implicit understanding that there is one, and only one, number label, either verbal or symbolic, during counting.

Opportunistic omnivore Any animal that eats a wide variety of foods, including fruits, vegetation, foliage, meat, etc. – anything that is non-toxic.

Pant-hoot Characteristic call given by chimpanzees under conditions of great excitement, such as greetings during reunions.

Parabolic receptors Body parts, usually ears or structures around the face, that allow for enhanced sound reception, particularly for vocalizations over long distances (e.g., large cheek pads of adult male orang-utans).

Perspective-taking The ability to mentally put one's self in the place of another, so that one "experiences" the world as the other individual does.

Pestle-pounding Tool use observed in one community of chimpanzees that involves moving to the top of a palm tree, pulling out the large centre crown and using the stalk as a pestle to pound and soften the core of tree, prior to eating it.

Playback study Experimental approach that uses pre-recorded sounds such as species-specific vocalizations that are then replayed to single subjects to observe their behavioural response.

Polygynous Type of social structure whereby a male mates with more than one female during the breeding season.

Precision grip The use of the thumb and index finger to pick up or hold a small object; relies upon an opposable thumb that is long and flexible enough to cross over the palm of the hand.

Prosimians The most primitive of non-human primates species; more dependent on smell than vision; mostly nocturnal (eg, bushbabies, tarsiers, lemurs, lorises).

Protoculture Rudimentary culture with some social traditions, including transmission of tool use, food sharing and cooperative living within a dynamic and complex social structure (eg, chimpanzees).

Rumble vocalizations Low frequency infrasound vocalizations produced by elephants for communication over long distance and detected through their feet.

Self-concept Mental perception of one's own being, existence and attributes.

Semi-arboreal Animal species that are active a portion of their time in the trees, while spending the majority of time on the ground.

Semi-terrestrial Animal species that are active a portion of their time on the ground, while spending the majority of time in the trees.

Sham-marking The act of pretending to place a mark on the face or body of an experimental subject, without leaving any visible result.

Social carnivore Any of the large cat species that live in complex social units, hunt cooperatively and raise their young communally.

Social learning Learning a behaviour or set of behaviours by observing others in one's social group that would otherwise not be acquired.

Theory of Mind (ToM) A learned understanding that other individuals have similar or different thoughts, and therefore minds and brains like our own; recognition that other's may have different knowledge states, beliefs or desires.

Trial-and-error learning Type of learning in which individuals attempt task solutions, sometimes making many errors over and over, until achieving task solution.

Win/stay; lose/shift learning Learning strategy whereby an animal taking part in a task comes to understand they must choose between two objects, giving them a 50/50 chance of being correct. If they are correct on their first choice during a two-choice task, the subject picks the same object on the following task (win/stay). If incorrect, subject should choose the other object (lose/shift).

INDEX

Page numbers in *italics* refer to illustration captions.

CREDITS

Quarto would like to thank the following agencies and research centres for kindly supplying images for inclusion in this book.

Key: T = top; B = bottom; L = left; R = right

Alain Darroux: 131
Alamy: 99
Amanda Pippin: 165
Amy Fultz (www.chimphaven.org): 155
Ardea: 11, 17, 18–19, 21, 68BL, 79, 90, 97, 114, 119, 175, 181, 183
Arlene Levin-Rowe (www.alexfoundation.org): 160T
Behavioural Ecology Research Group, Oxford: 22–23
Corbis: 2, 8–9, 139, 168–169
Diana Reiss: 123
FLPA: 110, 116, 176
Getty: 37, 177, 185
iStock
Nature Picture Library: 31, 53, 62, 68BR, 75, 81, 111, 106–107, 171
NHPA: 37, 48, 51, 69BR
Photolibrary: 3, 4, 6–7, 10, 12–13, 15, 24–25, 33, 36, 39, 41, 45, 47T, 49, 55, 57, 67, 69BL, 60–61, 71, 73, 77, 84–85, 87, 101, 103, 105, 121, 122, 133, 137, 141, 143, 151, 161B, 173, 179
Shutterstock
Tatyana Humle (www.pri.Kyoto–u.ac.jp): 35
Tetsuro Maetsuzawa (www.greenpassage.org): 163
The Great Ape Trust of Iowa (www.thegreatapetrust.org): 162
Wikipedia: 47B, 129